OCR

Updated

Chemistry

REVISION GUIDE

Sandra Clinton
Emma Poole

AS

OXFORD

Oxford University Press is a department of the University of Oxford. It furthers the University's objective of excellence in research, scholarship, and education by publishing worldwide in

Oxford New York Auckland Cape Town Dar es Salaam Hong Kong Karachi Kuala Lumpur Madrid Melbourne Mexico City Nairobi New Delhi Shanghai Taipei Toronto

With offices in Argentina Austria Brazil Chile Czech Republic France Greece Guatamala Hungary Italy Japan South Korea Poland Portugal Singapore Switzerland Thailand Turkey Ukraine Vietnam

British Library Cataloguing in Publication Data
Data available

ISBN: 978-0-19913628-5

10 9 8 7 6 5 4 3 2 1

Printed by Bell and Bain Ltd., Glasgow

Acknowledgements:
Authors, editors, co-ordinators and contributors: Sandra Clinton, Emma Poole, Ruth Holmes.
Project managed by Elektra Media Ltd. Typeset by Wearset Ltd.

Contents

WELCOME AND INTRODUCTION

Welcome to the Revision Guide for OCR AS Chemistry.

We've tried to package the course in such a way as to help you more easily go into the examination room with added confidence for success.

You'll find details on:

- how OCR assess you through the examinations together with how examinations work
- there's all the content from the course presented in Specification order (with a handy list of Specification references for easy location)
- revision guidance.

Sandra Clinton

Emma Poole

2011

AS GCE SCHEME OF ASSESSMENT

The table below shows how the marks are allocated in the OCR AS Chemistry course.

Unit	Method of assessment	% of AS
AS Unit F321: *Atoms, Bonds, and Groups*	1 hour written paper Candidates answer all questions. 60 marks	30%
AS Unit F322: *Chains, Energy, and Resources*	1 hour 45 minutes written paper Candidates answer all questions. 100 marks	50%
AS Unit F323: *Practical Skills In Chemistry 1*	Coursework Candidates complete three tasks set by the exam board and marked internally. 40 marks	20%

Assessment objectives

Candidates are expected to demonstrate the following in the context of the chemical content of AS Chemistry:

Knowledge and understanding

- recognise, recall and show understanding of scientific knowledge;
- select, organise and communicate relevant information in a variety of forms.

Application of knowledge and understanding

- analyse and evaluate scientific knowledge and processes;
- apply scientific knowledge and processes to unfamiliar situations including those related to issues;
- assess the validity, reliability and credibility of scientific information.

How science works

- demonstrate and describe ethical, safe and skilful practical techniques and processes, selecting appropriate qualitative and quantitative methods;
- make, record and communicate reliable and valid observations and measurements with appropriate precision and accuracy;
- analyse, interpret, explain and evaluate the methodology, results and impact of their own and others' experimental and investigative activities in a variety of ways.

Quality of written communication

It's all very well knowing lots of facts but you need to be able to communicate what you know and get your ideas across to the examiner. You might not think it but the examiner has your interests at the centre of their job – you need to give them the easiest route to maximizing your marks.

You should:

- ensure that text is legible and that spelling, punctuation, and grammar are accurate so that meaning is clear
- select and use a form and style of writing appropriate to purpose and to complex subject matter
- organise information clearly and coherently, using specialist vocabulary where appropriate

Quality of written communication is assessed across all externally assessed Units; if you write clear, well explained answers then you should obtain any marks assigned to it.

HOW EXAMINATIONS WORK

Part of your course involves understanding/knowing *how science works*. Well, for maximum marks in your exam you need to know *how examinations work*. And in much the same way as science there are in-built rules that form the foundation of how examinations are constructed.

Get and speak the *lingo*: know exam-speak, play the game. Here are some popular terms which are often used in exam questions. Make sure you know what each of these terms means. For a term that requires a written answer it is most unlikely that one/two words will do!

- **Calculate**: means calculate and write down the numerical answer to the question. Remember to include your working and the units.
- **Define**: write down what a chemical/term means. Remember to include any conditions involved.
- **Describe**: write down using words and, where appropriate diagrams, all the key points. Think about the number of marks that are available when you write your answer.
- **Discuss**: write down details of the points in the given topic.
- **Explain**: write down a supporting argument using your chemical knowledge. Think about the number of marks that are available when you write your answer.
- **List**: write down a number of points. Think about the number of points required. Remember, incorrect answers will cost you marks.
- **Sketch**: when this term is used a simple freehand answer is acceptable. Remember to make sure that you include any important labels.
- **State**: write down the answer. Remember a short answer rather than a long explanation is required.
- **Suggest**: use your chemical knowledge to answer the question. This term is used when there is more than one possible answer or when the question involves an unfamiliar context.

GETTING DOWN TO REVISION

OK, you're committed to preparing for your examination! How do *you* go about it? Remember, there are almost as many ways to revise as there are students revising. Underneath the methods of revising there are some common goals that the revising has to achieve.

1 Boost your confidence

Careful revision will enable you to perform at your best in your examinations. So give yourself the easiest route through the work. Organise the work into small, manageable chunks and set it out in a timetable. Then each time you finish a chunk you can say to yourself 'done it', and then move on to the next one. **And give yourself a reward too!** It's amazing how much of a lift it gives you by working in this way.

2 Be successful

To be successful in AS level chemistry you must be able to:

- recall information
- apply your knowledge to new and unfamiliar situations
- carry out precise and accurate experimental work*
- interpret and analyse both your own experimental data and that of others*

*experience gained with practical work will help you with answering questions in the examination so don't set aside all this valuable knowledge and understanding

How this revision guide can help

This book will provide you with the facts which you need to recall and some examination practice.

Use it as a working book. Start with the Contents list that shows you each of the topics covered. Print them and highlight the areas which you are already confident about. Then, in a different colour mark off the sections one at a time as your revision progresses. By doing this you will feel positive about what you have achieved rather than negative about what you still have to do.

For your revision programme you might like to use some or all of the following strategies:

- read through the topics one at a time and try the quick questions
- choose a topic and make your own condensed summary notes
- print or sketch then colour important diagrams
- highlight key definitions or write them onto flash cards
- work through facts until you can recall them
- constantly test your recall by covering up sections and writing them from memory
- ask your friends and family to test your recall
- make posters for your bedroom walls
- use the 'objectives' as a self-test list
- carry out exam practice
- work carefully through the material on each page
- make 'to do' lists and tick them off.

Whatever strategies you use, measure your revision in terms of the progress you are making rather than the length of time you have spent working. You will feel much more positive if you are able to say specific things you have achieved at the end of a day's revision rather than thinking, 'I spent eight hours inside on a sunny day!'. Don't sit for extended periods of time. Plan your day so that you have regular breaks, fresh air, and things to look forward too.

Watch out: revision is an active occupation - just reading information is not enough! You will need to be active in your work for your revision to be successful.

Improving your recall: A good strategy for recalling information is to focus on a small number of facts for five minutes. Copy out the facts repeatedly in silence for five minutes then turn your piece of paper over and write them from memory. If you get any wrong then just write these out for five minutes. Finally test your recall of all the facts. Come back to the same facts later in the day and test yourself again. Then revisit them the next day and again later in the week. By carrying out this process they will become part of your long term memory – you will have learnt them!

Past paper practice: Once you have built up a solid factual knowledge base you need to test it by completing past paper practice. It might be a good idea to tackle several questions on the same topic from a number of papers rather than working through a whole paper at once. This will enable you to identify any weak areas so that you can work on them in more detail. Finally, remember to complete some mock exam papers under exam conditions.

A final word (or two) for the examination room

Unlike the GCSE examination this examination does not have a foundation tier and a higher tier so you must be prepared to answer questions on all the topics outlined in the Specification. Here are some obvious, and not so obvious thoughts:

- read through all the questions (*obvious*)
- identify which questions you can answer well (*obvious*)
  - start by answering these questions (*not so obvious*)
- once you have read a question carefully make sure you answer that question and NOT something you might think is the question (*not so obvious*)
- look at the number of marks that are available for each question and take this into account when you write your answer (*obvious and not so obvious*)
- answer space: the amount of space left for the answer will give you an indication of the length of answer the examiner expects (*obvious*)
  - short space = short answer (*obvious*)
  - longer space(1) = extended answer probably required; perhaps a sentence, or two; a calculation with working; a list containing a selection items (*obvious*)
  - longer space(2) = one word answer unlikely to be sufficient (*not so obvious*)
- answer all the questions, even if you have to guess at some (*obvious*)
- pace yourself and try to leave enough time to check your answers at the end (*obvious*)

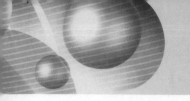

1.01 Atomic structure

Atomic number, mass number and isotopes

Atomic number (Z): The number of protons in the nucleus.

Mass number (A): The number of protons and neutrons in the nucleus.

Isotope: Atoms that have the same number of protons but different numbers of neutrons in their nucleus.

Atoms and sub-atomic particles

Atoms are made from sub-atomic or fundamental particles called **protons**, **neutrons**, and **electrons**. The masses and charges of these fundamental particles are very small so we look at them relative to the proton.

particle	relative mass	relative charge
proton	1	+1
neutron	1	0
electron	1/2000	−1

The mass of an electron is so small it is often considered to be negligible. Protons and neutrons are located in the nucleus of the atom while electrons orbit the nucleus.

Atomic number and mass number

Ensure that you can recall the definitions of atomic number and mass number.

- Atoms can be represented: $_Z^A X$
- Number of protons = atomic number (Z)
- Number of electrons = atomic number (Z) (ONLY for neutral atoms)
- Number of neutrons = mass number – atomic number ($A - Z$)

For ions (particles which have lost or gained electrons) the charge needs to be taken into account.

Beryllium atoms $_4^9 Be$ protons = 4, electrons = 4, neutrons = 9 − 4 = 5

Fluoride ions $_9^{19} F^-$ protons = 9, electrons = 9 + 1 = 10, neutrons = 19 − 9 = 10

Calcium ions $_{20}^{40} Ca^{2+}$ protons = 20, electrons = 20 − 2 = 18, neutrons = 40 − 20 = 20

Isotopes

Make sure that you can remember the definition of an **isotope**.

- Isotopes contain differing numbers of neutrons but the same number of protons.
  - As a result they are chemically identical.

There are two isotopes of chlorine

$_{17}^{37} Cl$ protons = 17, electrons = 17, neutrons = 20
$_{17}^{35} Cl$ protons = 17, electrons = 17, neutrons = 18

Early models of the atom

The model of the atom has been developed over many hundreds of years. Scientists are constantly working to improve this model.

Four elements exist 'Fire, Earth, Air, water' **Aristotle**	Discovered electrons. Proposed the plum pudding model of atoms. **JJ Thomson**	Linked atomic number to the number of protons in the nucleus. **Moseley**	

470-380 BCE	384-322 BCE	1600s	1897	1911	1913	1932

Leucippus and Democritus Early theory of the atom. All matter is made of atoms that are too small too see and cannot be cut up.	**Dalton then Boyle** Returned to and developed the original theory. Elements contain identical atoms. Atoms do not change in chemical reactions.	**Rutherford with Geiger and Marsden** Proved that atoms had a nucleus containing protons. Electrons orbit the nucleus.	**Chadwick** Discovered neutrons.

Relative atomic mass, A_r

The mass of all atoms is measured relative to the mass of a ^{12}C atom. Relative atomic mass can be calculated using percentage data for an element:

isotope	204	206	207	208
percentage	10	20	20	50

- the A_r of lead can then be calculated as a weighted mean average.

$$A_r = \frac{(204 \times 10) + (206 \times 20) + (207 \times 20) + (208 \times 50)}{100} = 207$$

Determining relative molecular mass and relative formula mass

Relative molecular mass is readily calculated from relative atomic mass values. Make sure you can follow the two examples below.

Calculate the relative molecular mass of carbon dioxide, CO_2.

- Look up the A_r values: carbon (12.0), oxygen (16.0).
- Add up the A_r values: $12.0 + (2 \times 16.0) = 44.0$.

Calculate the relative molecular mass of ethanol, C_2H_5OH.

- Look up the A_r values: carbon (12.0), hydrogen (1.0), oxygen (16.0).
- Add up the A_r values: $(2 \times 12.0) + (6 \times 1.0) + 16.0 = 46.0$.

Always carry out your calculations using the same degree of accuracy as given on the data sheet (1 decimal place).

Relative formula mass is used for ionic substances, where we calculate the mass of one formula unit. You might want to come back to this section when you have revised ionic bonding and structure.

Calculate the relative formula mass of calcium carbonate, $CaCO_3$.

- Look up the A_r values: carbon (12.0), calcium (40.1), oxygen (16.0).
- Add up the A_r values: $40.1 + 12.0 + (3 \times 16.0) = 100.1$.

Definitions

Relative atomic mass

A_r = weighted mean mass of an atom of an element compared with one-twelfth of the mass of an atom of carbon-12.

Relative molecular mass

M_r = weighted mean mass of an atom of a molecule compared with one-twelfth of the mass of an atom of carbon-12.

Relative formula mass

Weighted mean mass of a formula unit compared with one-twelfth of the mass of an atom of carbon-12.

Relative isotopic mass

The mass of an atom of an isotope compared with one-twelfth of the mass of an atom of carbon-12.

Questions

1 Write down the numbers of protons, neutrons and electrons in the following:

 a $^{23}_{11}Na$ b $^{16}_{8}O^{2-}$ c $^{80}_{35}Br$ d $^{40}_{19}K^+$

2 Use the percentage by mass data given below to determine the relative atomic mass and identity of the element below.

isotope	20	21	22
percentage	90.48	0.27	9.25

3 Calculate the relative molecular mass of:

 a methane, CH_4

 b ethene, C_2H_4

 c water, H_2O

4 Calculate the relative formula mass of:

 a magnesium oxide, MgO

 b sodium sulfate, Na_2SO_4

 c magnesium hydroxide, $Mg(OH)_2$

Shapes of orbitals

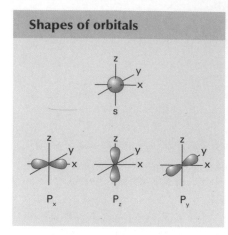

P_x P_z P_y

Configuration examples

- Potassium has 19 electrons. Its electron configuration is: $1s^2\ 2s^2\ 2p^6\ 3s^2\ 3p^6\ 4s^1$
- Scandium has 21 electrons. Its electron configuration is: $1s^2\ 2s^2\ 2p^6\ 3s^2\ 3p^6\ 4s^2\ 3d^1$
- A noble gas core is written in place of the inner shell electrons. This is shown using square brackets.
- Vanadium would be written $[Ar]\ 4s^2\ 3d^3$ since the electron configuration of argon is $1s^2\ 2s^2\ 2p^6\ 3s^2\ 3p^6$.
- Sodium could be written as $[Ne]\ 3s^1$.
- **Exception 1:** chromium has the configuration $[Ar]\ 3d^5\ 4s^1$ which is a more stable than $[Ar]\ 3d^4\ 4s^2$.
- **Exception 2:** copper has the configuration $[Ar]\ 3d^{10}\ 4s^1$ which is more stable than $[Ar]\ 3d^9\ 4s^2$.

Energy levels

Electrons are in constant motion around the nucleus of an atom.

- Electrons are found in **energy levels** represented by a quantum number.
- These are split into **sub-levels** with different maximum numbers of electrons.
- Different types of sub-level contain different numbers of **orbitals**.
- Each orbital can hold two electrons (one spinning up and one spinning down).
- Electrons are constantly moving so their exact location cannot be identified.
- The shape of an orbital tells you where an electron is most likely to be found.

energy level	1	2		3			4			
type of sub-level	s	s	p	s	p	d	s	p	d	f
number of orbitals in sub-level	1	1	3	1	3	5	1	3	5	7
maximum number of electrons in sub-level	2	2	6	2	6	10	2	6	10	14
maximum number of electrons in level	2	8		18			32			

Electron configurations

Use three principles for writing electron configurations. Work through each rule considering the examples shown in the table below. You won't be expected to name the rules in the examination but make sure you can apply each of them.

The 4s orbital is slightly lower in energy than the 3d so is filled first.

- The Aufbau principle – electrons fill energy levels in order of increasing energy.
- The Pauli exclusion principle – each orbital can contain a maximum of two electrons. Electrons in the same orbital must have opposite spins. This is shown using an up arrow and a down arrow.
- Hund's rule – orbitals are occupied unpaired before paired.

Element	Electron configuration	Spin diagrams								
		1s	2s	2p			3s	3p		
H	$1s^1$	↑								
He	$1s^2$	↑↓								
Li	$1s^2\ 2s^1$	↑↓	↑							
C	$1s^2\ 2s^2\ 2p^2$	↑↓	↑↓	↑	↑					
O	$1s^2\ 2s^2\ 2p^4$	↑↓	↑↓	↑↓	↑	↑				
P	$1s^2\ 2s^2\ 2p^6\ 3s^2\ 3p^3$	↑↓	↑↓	↑↓	↑↓	↑↓	↑↓	↑	↑	↑

- The period that an element is in determines its highest energy level.
- The group that an element is in determines its outer electron configuration.

element	period	highest energy level	block and group	outer electron configuration	electron configuration
Na	3	3	s block group 1	s^1	$1s^2\ 2s^2\ 2p^6\ 3s^1$
N	2	2	p block group 5	s^2p^3	$1s^2\ 2s^2\ 2p^3$
V	3	3	d block	d^3	$1s^2\ 2s^2\ 2p^6\ 3s^2\ 3p^6\ 4s^2\ 3d^3$

Ionization energy

The values obtained for ionization energies provide substantial evidence for the existence of energy levels and sub-levels. Make sure you memorize the definition for ionization energy.

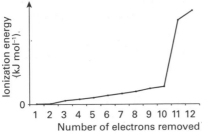

Successive ionization energies of magnesium

First ionization energies of group 2 elements

Ionization energy is influenced by

- the number of protons in the nucleus of an atom
- the distance of the outermost electron from the nucleus (the further it is from the nucleus the more the outer electron is shielded from the effect of the nucleus)
  - As a result the first ionization energy decreases as you go down a group.

Ionization energy

First ionization energy is the energy required to remove 1 electron from each atom in 1 mole of gaseous atoms forming 1 mole of ions with a single positive charge.

e.g. The equation for the first ionization (of magnesium) is:

$$Mg(g) \rightarrow Mg^+(g) + e^-$$

Successive ionization energies

Successive ionization energies represent the energy required to remove 1 electron from each ion in 1 mole of gaseous $^{n+}$ ions to form one mole of $^{(n+1)+}$ ions.

For example:

The equation for the third ionization of sulfur is:

$$S^{2+}(g) \rightarrow S^{3+}(g) + e^-$$

Successive ionization energies of magnesium

Successive ionization energies provide further evidence for the existence of energy levels.

- There is a general increase in the energy needed to remove each electron from magnesium.
- This is because the electron is being removed from an ion with an increasing positive charge.
- There is a very big increase between the tenth and eleventh ionization energies as the eleventh electron is being removed from a full orbital that is nearer to the nucleus and of lower energy than the tenth electron.
- The group to which an element belongs can be determined by identifying where the big jump in ionization energy occurs.

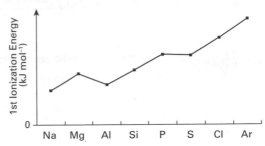

First ionization energies of the elements in period 3.

First ionization energies of the elements in period 3

The graph of first ionization energies for period 3 elements provides substantial support for the existence of energy sub-levels.

- The number of protons increases across period 3 so there is an increase in the charge on the nucleus.
  - As a result the force of attraction between the nucleus and the outer electron increases.
- The number of electrons also increases but these go into the same energy level so are at the same distance from the nucleus and experience the same shielding.
  - As a result there is an increase in first ionization energy across period 3.

Ions of d block elements

- Iron has the configuration [Ar] $3d^6$ $4s^2$: its ions, Fe^{2+} and Fe^{3+} are [Ar] $3d^6$ and [Ar] $3d^5$.

Electron configurations of ions

element	lithium	aluminium
group	1	3
electron configuration of atom	$1s^2$ $2s^1$	$1s^2$ $2s^2$ $2p^6$ $3s^2$ $3p^1$
electron configuration of ion	$1s^2$	$1s^2$ $2s^2$ $2p^6$
symbol of ion	Li^+	Al^{3+}
element	oxygen	fluorine
group	6	7
electron configuration of atom	$1s^2$ $2s^2$ $2p^4$	$1s^2$ $2s^2$ $2p^5$
electron configuration of ion	$1s^2$ $2s^2$ $2p^6$	$1s^2$ $2s^2$ $2p^6$
symbol of ion	O^{2-}	F^-

Questions

1 Draw out an electron spin diagram and then write electron configurations using a noble gas abbreviation for:

   a Na   **b** S   **c** Cl   **d** Be

2 Write full electron configurations for a:

   a magnesium ion   **b** sulfide ion
   c oxide ion   **d** sodium ion

3 Sketch a graph showing the successive ionization energies of oxygen. Label the graph in detail, explaining the patterns it shows.

Moles and the Avogadro constant

One mole is the amount of a substance that contains the same number of particles as there are atoms in exactly 12 g of ^{12}C.

The Avogadro constant (6.02×10^{23}) is the number of particles in one mole of a substance.

Amedeo Avogadro was an Italian scientist whose work was not accepted until after his death.

Molar mass

The mass per mole of a substance, units of $g\,mol^{-1}$. It has the symbol M.

Amount of substance

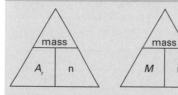

A_r = relative atomic mass

M = molar mass / formula mass

n = number of moles

n = number of moles

Empirical formulae and molecular formulae

The empirical formula shows the simplest whole number ratio of the atoms of each element in a compound.

The molecular formula shows the actual number of atoms of each element in one molecule of the compound. The molecular formula is a whole number multiple of the empirical formula.

The mole and the Avogadro constant

Chemists are interested in how many atoms, molecules, ions, etc. take part in reactions. All of these particles are very small so we cannot determine their mass. Instead we carry out calculations using the concept of the **mole**.

- The mole is the unit for amount of substance.
- One mole of a substance contains the same number of particles as there are atoms in exactly 12 g of ^{12}C.
- The number of particles in one mole of a substance is the **Avogadro constant** N_A (after Amedeo Avogadro), $6.02 \times 10^{23}\,mol^{-1}$.
- Molar mass is the mass per mole of a substance, units $g\,mol^{-1}$.

Amount of substance

The amount of a substance is readily calculated from the relative atomic mass or the molar mass of a substance using the relationship:

mass = $M \times n$

You must be able to manipulate this relationship in order to calculate mass, M, or n given suitable data.

Calculations using mass = $M \times n$

a Calculate the mass in grams of 2 moles of Mg atoms.

$$\text{mass} = A_r \times n = 24.3 \times 2 = 48.6 \text{ g}$$

b Calculate the mass in grams of 0.74 moles of $Ca(OH)_2$.

$$\text{mass} = M \times n = [40.1 + 2(17)] \times 0.74 = 74.1 \times 0.74 = 54.8 \text{ g}$$

c Calculate the number of moles of Na atoms in 6.5 g of Na atoms.

$$n = \text{mass}/A_r = 6.5/23.1 = 0.28$$

d Calculate the number of moles of MgO in 2×10^{-6} g of MgO.

$$n = \text{mass}/M = 2 \times 10^{-6}/40.3 = 4.96 \times 10^{-8}$$

e Calculate the M of aluminium oxide, Al_2O_3, if 0.5 moles of aluminium oxide has a mass of 51 g

$$M = \text{mass}/n = 51/0.5 = 102 \text{ g mol}^{-1}$$

Empirical and molecular formulae

The **empirical formula** of a compound is the simplest whole number ratio of the atoms of the elements in a compound. The **molecular formula** is the actual number of atoms.

For example: The empirical formula of butene is CH_2

The molecular formula of butene is C_4H_8

The empirical formula of a compound can be calculated if the percentage composition of the compound is known.

Calculating empirical formula

A compound contains 40.06% calcium, 11.99% carbon, and 47.95% oxygen by mass. Determine its empirical formula.

It is essential to show all your working in these calculations.

element	Ca	C	O
% by mass	40.06	11.99	47.95
$\div A_r$	0.999	0.999	2.997
$\div$ smallest	1	1	3

Empirical formula: $CaCO_3$

Gases

Under standard conditions 1 mole of any gas occupies 24 dm^3.
The volume of a gas is readily calculated using the relationship:

$$\text{vol} = n \times 24$$

Remember to convert gas volumes into dm^3 when carrying out calculations.

Molar volume

The volume occupied by 1 mole of gas, 24 dm^3 mol^{-1} under standard conditions.

Calculations using vol = n × 24

Carry out the following calculations giving your final answers to 2 significant figures:

a Calculate the number of moles in 46 dm^3 of hydrogen gas.

$$\text{moles} = \frac{46}{24} = 1.9$$

b Calculate the number of moles in 250 cm^3 of oxygen gas.

$$\text{moles} = \frac{(250/1000)}{24} = 0.010$$

c Calculate the volume of 3.6 moles of carbon dioxide gas.

$$\text{volume} = 3.6 \times 24 = 86 \text{ dm}^3$$

d Calculate the mass of 100 cm^3 of methane gas.

$$\text{moles} = \frac{(100/1000)}{24} = 4.167 \times 10^{-3}$$
$$\text{mass} = 4.167 \times 10^{-3} \times 16 = 0.067 \text{ g}$$

Amount of substance

n = number of moles

vol = volume of gas in dm^3

Standard conditions

Pressure of 100 kPa (1 atmosphere).

Temperature of 298 K (25 °C).

Concentration of 1 mol dm^{-3} (for aqueous solutions).

Calculations from equations

2.0 g of magnesium are fully reacted with hydrochloric acid forming magnesium chloride and hydrogen gas. Calculate the volume of hydrogen gas released (in dm^3).

Step 1 Write a balanced equation.

$$Mg(s) + 2HCl(aq) \rightarrow MgCl_2(aq) + H_2(g)$$

Step 2 Calculate the number of moles of the substance you have the most information about.

$$\text{number of moles of Mg} = \frac{2.0}{24.3} = 0.0823$$

Step 3 Use the balanced equation to calculate the number of moles of the unknown substance.

number of moles of hydrogen = 0.0823

Step 4 Convert the number of moles to the correct unit.

volume of hydrogen gas = 0.0823 × 24 = 1.98 dm^3

When methane is fully combusted in oxygen it forms carbon dioxide gas and water. What mass of methane is burnt in order to obtain 2.5 dm^3 of carbon dioxide? Give your answer to 2 decimal places.

Step 1 Write a balanced equation.

$$CH_4(g) + 2O_2(g) \rightarrow CO_2(g) + 2H_2O(l)$$

Step 2 Calculate the number of moles of the substance you have the most information about.

$$\text{number of moles of CO}_2 = \frac{2.5}{24} = 0.104$$

Step 3 Use the balanced equation to calculate the number of moles of the unknown substance.

number of moles of methane = 0.104

Step 4 Convert the number of moles into the correct unit.

mass of methane = 0.104 × 16 = 1.67 g

Questions

1 Calculate the following. Remember to show all your working.

  a the number of moles of CaO in 4 g of CaO

  b the mass of 0.15 moles of K

  c the mass of 0.32 moles of LiOH

2 Calculate the following. Show all your working.

  a the number of moles of methane gas, CH$_4$, in 2.8 dm^3 of CH$_4$

  b the volume occupied by 3 moles of oxygen gas

  c the number of moles of hydrogen gas, H$_2$, in 2758 cm^3 of H$_2$

3 Using the equation below calculate the volume of carbon dioxide produced (in cm^3) when 2 g of calcium carbonate react fully with sulfuric acid.

$$CaCO_3(s) + H_2SO_4(aq) \rightarrow CaSO_4(aq) + CO_2(g) + H_2O(l)$$

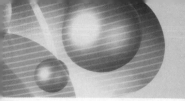

1.04 Moles and calculations: solutions

Solutions

Concentration is the amount of a substance in moles dissolved in 1 dm³ of water.

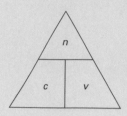

n = Number of moles
c = concentration in mol dm⁻³
v = volume of solution in dm³

Concentration

Solutions of a high concentration are described as **concentrated**.

Solutions of low concentration are described as **dilute**.

Concentrations

A great deal of chemistry is carried out in aqueous solution. So a good understanding of the term concentration is essential for success at AS level.

- **Concentration** is the amount of a substance in moles dissolved in 1 dm³ of water (1 dm³ = 1000 cm³).
- This means that a 2 mol dm⁻³ solution of sulfuric acid contains 2 moles of sulfuric acid dissolved in 1 dm³ of water.
- This is 2 × 98.1 = 196.2 g of sulfuric acid.

The relationship between concentration and number of moles is: $n = c \times v$

n = number of moles, c = concentration in dm³, v = volume of solution in dm³

Note – The volume used in the concentration expression is needed in dm³. To convert dm³ to cm³ you must multiply by 1000. To convert cm³ to dm³ you must divide by 1000.

Acid–base titrations

Chemists use **titrations** to find the concentration of a reactant in solution.

At As level you only need to know about acid-base titrations.

- A solution of known concentration (a standard solution) is reacted with the solution of unknown concentration.
- Very precise apparatus (burettes and pipettes) is used to measure the volumes of solutions.
- An indicator is used to enable the point of exact neutralization to be determined.

Remember that reactions take place in molar proportions so it is essential you write a balanced equation to help you determine mole ratios.

For example:

hydrochloric acid and sodium hydroxide react in a 1:1 ratio

$$HCl(aq) + NaOH(aq) \rightarrow NaCl(aq) + H_2O$$

sulfuric acid and sodium hydroxide react in a 1:2 ratio

$$H_2SO_4(aq) + 2NaOH(aq) \rightarrow Na_2SO_4(aq) + 2H_2O$$

Acid–base titration calculations

Follow these systematic steps to carry out a titration calculation. In an examination question you are likely to be guided through some of these steps. Make sure you practise lots of these!

25.0 cm³ of sodium hydroxide solution is exactly neutralized by 21.40 cm³ of 0.5 mol dm⁻³ hydrochloric acid. What was the concentration of the sodium hydroxide solution?

Step 1 Write a balanced equation
$NaOH(aq) + HCl(aq) \rightarrow NaCl(aq) + H_2O(l)$

Step 2 Write out the data given in the question under the equation.
$NaOH(aq)$ + $HCl(aq)$ → $NaCl(aq)$ + $H_2O(l)$
25.0 cm³   21.40 cm³
?   0.5 mol dm⁻³

Step 3 Convert the substance you know the most about into moles.
number of moles of HCl(aq):
$n = c \times v = 0.5 \times (21.4 \div 1000) = 0.0107$

Step 4 Use the balanced equation to determine the moles of the unknown substance.
1 mol HCl reacts with 1 mol NaOH.
So number of moles of NaOH = 0.0107

Step 5 Convert the number into the units asked for in the question.
concentration of sodium hydroxide solution:
$c = n/v = 0.0107/(25.0 \div 1000) = 0.428$ mol dm⁻³

More challenging calculations

Work carefully through the sample calculations below. You will be guided through some of these steps in the exam.

1 25.0 cm³ of sodium hydroxide are exactly neutralized by 13.0 cm³ of 0.1 mol dm⁻³ sulfuric acid. What was the concentration of the sodium hydroxide solution?

Step 1 Write a balanced equation.
$2NaOH(aq) + H_2SO_4(aq) \rightarrow Na_2SO_4(aq) + 2H_2O(l)$

Step 2 Write out the data given in the question under the equation.
$2NaOH(aq) + H_2SO_4(aq) \rightarrow Na_2SO_4(aq) + 2H_2O(l)$
25.0 cm³   13.0 cm³
   ?   0.1 mol dm⁻³

Step 3 Convert the substance you know the most about into moles.
Number of moles of $H_2SO_4(aq)$:
$n = c \times v = 0.1 \times (13/1000) = 1.3 \times 10^{-3}$

Step 4 Use the balanced equation to determine the moles of the unknown substance.
1 mol H_2SO_4 reacts with 2 mol NaOH.
So number of moles of NaOH = $1.3 \times 10^{-3} \times 2 = 2.6 \times 10^{-3}$

Step 5 Convert the number into the units asked for in the question.
Concentration of sodium hydroxide solution:
$c = \dfrac{n}{v} = \dfrac{(2.6 \times 10^{-3})}{(25.0 \div 1000)} = 0.104$ mol dm⁻³

2 25.0 cm³ of barium hydroxide of concentration 0.1 mol dm⁻³ are exactly neutralized by 40.0 cm³ of hydrochloric acid. What was the concentration of the hydrochloric acid solution?

Step 1 Write a balanced equation.
$Ba(OH)_2(aq) + 2HCl(aq) \rightarrow BaCl_2(aq) + 2H_2O(l)$

Step 2 Write out the data given in the question under the equation.
$Ba(OH)_2(aq) + 2HCl(aq) \rightarrow BaCl_2(aq) + 2H_2O(l)$
25.0 cm³   40.0 cm³
0.1 mol dm⁻³   ?

Step 3 Convert the substance you know the most about into moles.
Number of moles of $Ba(OH)_2(aq)$:
$n = c \times v = 0.1 \times (25.0/1000) = 2.5 \times 10^{-3}$

Step 4 Use the balanced equation to determine the moles of the unknown substance.
1 mol $Ba(OH)_2$ reacts with 2 mol HCl.
So number of moles of HCl = $2.5 \times 10^{-3} \times 2 = 5.0 \times 10^{-3}$

Step 5 Convert the number into the units asked for in the question.
Concentration of hydrochloric acid solution:
$c = \dfrac{n}{v} = \dfrac{(5.0 \times 10^{-3})}{(40.0 \div 1000)} = 0.125$ mol dm⁻³

Questions

1 Calculate the following. Remember to show all your working.

  a the concentration of a solution made by dissolving 6 g of NaOH in 1 dm³ of water

  b the concentration of a solution made by dissolving 3.2 g of HCl in 300 cm³ of water (note you need to use two of the mole relationships here)

  c the number of moles of H_2SO_4 present in 25.0 cm³ of 0.1 mol dm⁻³ $H_2SO_4(aq)$

  d the number of moles of HNO_3 present in 21.7 cm³ of 0.2 mol dm⁻³ $HNO_3(aq)$

  e the volume of 2 mol dm⁻³ KOH needed to provide 0.25 mol of KOH

2 25.0 cm³ of NaOH(aq) is exactly neutralized by 27.60 cm³ of 0.15 mol dm⁻³ $HNO_3(aq)$. Calculate the concentration of the NaOH(aq) solution.

3 25.0 cm³ of KOH(aq) are exactly neutralized by 26.00 cm³ of sulfuric acid of concentration 0.12 mol dm⁻³. Calculate the concentration of the KOH(aq) solution.

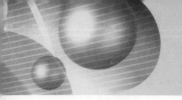

1.05 Writing and balancing equations

Combustion reactions

All substances when burnt in air (or pure oxygen) form oxides.

e.g. calcium + oxygen → calcium oxide

Hydrocarbons when burnt in unlimited oxygen form carbon dioxide and water.

e.g. methane + oxygen → carbon dioxide + water

Acid reactions

Acids are proton donors. In their reactions their H^+ ion is displaced and a salt is formed.

You should be familiar with the following acid reactions:

acid + metal → salt + hydrogen

acid + metal oxide → salt + water

acid + metal carbonate → salt + water + carbon dioxide

acid + alkali → salt + water

Balanced chemical equations

In a chemical reaction, reactants are changed into products. **Balanced equations** are written to show the reactants and products for reactions. It is important to remember that the number of each type of atom on the left hand side of the equation must be the same as the number of the same type of atom on the right hand side.

- You must follow the same steps each time you write an equation for a reaction.
- There is no need to write anything down except the final equation but you must think through the process in a logical way.

Writing a balanced equation

We can use the reaction of magnesium with oxygen as an example.

Step 1 Work out the identities of the reactants and products

$$magnesium + oxygen → magnesium\ oxide$$

Step 2 Construct formulae for each of the reactants and products

$$Mg + O_2 → MgO$$

Step 3 Balance the equation so that the number of each type of atom is the same on each side of the equation. Work from left to right balancing each atom in turn.

$$\mathbf{2}Mg + O_2 → \mathbf{2}MgO$$

Step 4 Add state symbols using (s) for solid, (l) for liquid, (g) for gas and (aq) for aqueous, a solution in water.

$$2Mg\mathbf{(s)} + O_2\mathbf{(g)} → 2MgO\mathbf{(s)}$$

In Unit 1 you will be expected to write and balance simple equations for reactions that you have studied in the unit. Examples include acid–base neutralization reactions that you carried out when completing titrations and equations for the acid reactions. You may also be given an unfamiliar equation and be asked to balance it. Do practise writing equations as you work through this revision book.

Balancing an equation for an unfamiliar reaction

Ammonia, NH_3, reacts with sodium to form sodium amide, $NaNH_2$, and hydrogen. Write a balanced equation for this reaction.

Step 1 Write formulae for the reactants on the left of an equation and for the products on the right.

$$NH_3 + Na → NaNH_2 + H_2$$

Step 2 Work through the equation from left to right balancing each atom in turn. There is one nitrogen atom on the left and one on the right so this equation balances in terms of nitrogen.

There are three hydrogen atoms on the left and four on the right. In order to provide enough hydrogen atoms for the right we must place a two in front of the ammonia.

$$\mathbf{2}NH_3 + Na → NaNH_2 + H_2$$

We now need to check the nitrogen atoms again; there are two on the left so we need to place a two in front of the sodium amide.

$$\mathbf{2}NH_3 + Na → \mathbf{2}NaNH_2 + H_2$$

The nitrogen and hydrogen atoms are now balanced. To finish the equation we need to place a two in front of the sodium atom.

$$\mathbf{2}NH_3 + \mathbf{2}Na → \mathbf{2}NaNH_2 + H_2$$

Ionic equations

Ionic equations can be written for precipitation, displacement, and neutralization reactions. Ions which play no part in the reaction are omitted. These are called spectator ions. For the reaction of hydrochloric acid and potassium hydroxide it is possible to write both a full equation and an ionic equation.

The full equation is: $HCl(aq) + KOH(aq) \rightarrow KCl(aq) + H_2O(l)$

In this equation note that the only substance which exists as a molecule is the water; all other species are ionic. To construct an ionic equation all the ions are written out, then those that are spectator ions can be cancelled out.

$$H^+(aq) + \cancel{Cl^-(aq)} + \cancel{K^+(aq)} + OH^-(aq) \longrightarrow \cancel{K^+(aq)} + \cancel{Cl^-(aq)} + H_2O(l)$$

This gives an ionic equation of $H^+(aq) + OH^-(aq) \rightarrow H_2O(l)$

Precipitation reactions are those in which a solid substance is formed by the combination of two ions in solution. These reactions are often used to test for the presence of ions. For example, the addition of silver nitrate solution to potassium iodide solution results in the formation of a yellow precipitate of silver iodide.

The full equation is: $AgNO_3(aq) + KI(aq) \rightarrow KNO_3(aq) + AgI(s)$

Both potassium and nitrate ions remain in solution so are spectator ions.

The ionic equation is $Ag^+(aq) + I^-(aq) \rightarrow AgI(s)$

Practising writing equations

Work through the following questions, cover up the answers then check them.

1 Balance the following equations. All individual formulae are correct.

  a $Li(s) + O_2(g) \circledR Li_2O(s)$

  b $N_2(g) + H_2(g) \circledR NH_3(g)$

  c $CH_4(g) + O_2(g) \circledR CO_2(g) + H_2O(l)$

2 Write equations for the following reactions.

  a Ethane gas, C_2H_6, reacts with oxygen gas forming carbon dioxide gas and steam.

  b Nitrogen gas reacts with oxygen gas forming nitrogen dioxide gas.

  c Solid calcium and water form calcium hydroxide solution and hydrogen gas.

Answers

1 a $4Li(s) + O_2(g) \circledR 2Li_2O(s)$

  b $N_2(g) + 3H_2(g) \circledR 2NH_3(g)$

  c $CH_4(g) + 2O_2(g) \circledR CO_2(g) + 2H_2O(l)$

2 a $2C_2H_6(g) + 7O_2(g) \circledR 4CO_2(g) + 6H_2O(l)$

  b $N_2(g) + 2O_2(g) \circledR 2NO_2(g)$

  c $Ca(s) + 2H_2O(l) \circledR Ca(OH)_2(aq) + H_2(g)$

Questions

1 Balance the following equations.

  a $Ba(s) + O_2(g) \rightarrow BaO(s)$

  b $Mg(s) + HCl(aq) \rightarrow MgCl_2(aq) + H_2(g)$

  c $K(s) + H_2O(l) \rightarrow KOH(aq) + H_2(g)$

2 Write ionic equations for the following reactions.

  a $HCl(aq) + NaOH(aq) \rightarrow NaCl(aq) + H_2O(l)$

  b $BaCl_2(aq) + MgSO_4(aq) \rightarrow BaSO_4(aq) + MgCl_2(aq)$

3 Methane, CH_4, and steam react forming carbon monoxide gas and hydrogen gas.

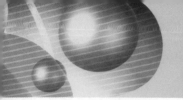

1.06 Acids, bases and salts

Acids

Acids release H^+ ions in aqueous solution.

Acids

Acids are substances that release H^+ ions when dissolved in water. As an H^+ ion is a proton acids may be referred to as proton donors.

Acids you will meet in AS include:

- hydrochloric acid $HCl(aq)$
- sulfuric acid $H_2SO_4(aq)$
- nitric acid $HNO_3(aq)$
- ethanoic acid $CH_3COOH(aq)$

In A2 you will lean about the difference between strong and weak acids.

Bases and alkalis

Bases accept H^+ ions.

Alkalis are soluble bases that release OH^- ions in aqueous solution.

Bases and alkalis

Bases are substances that accept H^+ ions. They may be referred to as proton acceptors.

Common bases include:

- metal oxides   e.g. magnesium oxide, MgO
- metal hydroxides   e.g. sodium hydroxide, $NaOH$
- ammonia NH_3

Alkalis are bases that dissolve in water releasing OH^- ions.

Common alkalis include:

- sodium hydroxide $NaOH$
- potassium hydroxide KOH
- aqueous ammonia NH_4OH

Salts

Salts are formed when the H^+ ion of an acid is replaced by a metal ion or by the ammonium ion, NH_4^+.

You must be able to write equations for the reaction of an acid with carbonates, bases, and alkalis. Work through the examples below. Try covering up the products in the reaction and completing them yourself before checking your answer.

acid + carbonate → salt + water + carbon dioxide
$$2HCl(aq) + CaCO_3(s) \rightarrow CaCl_2(aq) + H_2O(l) + CO_2(g)$$
$$H_2SO_4(aq) + Na_2CO_3(aq) \rightarrow Na_2SO_4(aq) + H_2O(l) + CO_2(g)$$

acid + base → salt + water
$$2HCl(aq) + MgO(s) \rightarrow MgCl_2(aq) + H_2O(l)$$

acid + alkali → salt + water
$$HNO_3(aq) + NaOH(aq) \rightarrow NaNO_3(aq) + H_2O(l)$$
$$H_2SO_4(aq) + 2KOH(aq) \rightarrow K_2SO_4(aq) + 2H_2O(l)$$

Note that in both of these reactions the OH^- ion from the alkali has accepted an H^+ ion forming H_2O.

Reactions with ammonia

Ammonia, NH_3, reacts with acids, accepting an H^+ ion and forming ammonium salts containing the ammonium ion, NH_4^+.

Work through the examples below carefully – writing equations for the reactions of ammonia needs careful thought.

hydrogen chloride + ammonia → ammonium chloride
$$HCl(g) + NH_3(g) \rightarrow NH_4Cl(s)$$

In aqueous solution the ammonia forms ammonium hydroxide, $NH_4OH(aq)$, which can release OH^- ions and is therefore an alkali.

hydrochloric acid + ammonia solution → ammonium chloride + water
$$HCl(aq) + NH_4OH(aq) \rightarrow NH_4Cl(aq) + H_2O(l)$$

sulfuric acid + ammonia solution → ammonium sulfate + water
$$H_2SO_4(aq) + 2NH_4OH(aq) \rightarrow (NH_4)_2SO_4(aq) + 2H_2O(l)$$

Anhydrous or hydrated?

Anhydrous substances contain no water molecules.

Hydrated substances are crystalline substances containing water molecules.

Water of crystallization is the water molecules that form part of the crystalline structure of a compound.

For example:

$CuSO_4(s)$ anhydrous copper sulfate

$CuSO_4.5H_2O(s)$ hydrated copper sulfate

The hydrated copper sulfate contains 5 moles of water of crystallization per mole of hydrated solid.

Calculating the formula of hydrated salts

Hydrated salts are crystalline solids containing water of crystallization. For example, hydrated copper chloride is $CuCl_2.2H_2O$.

The formula of a hydrated salt can be calculated from different types of data.

Worked example 1

3.00 g of hydrated copper sulfate are found to contain 1.92 g of copper sulfate and 1.08 g of water. Determine the formula of the hydrated copper sulfate.

Step 1 Calculate the moles of each substance.

	$CuSO_4$	H_2O
mass in g	1.92	1.08
moles	$\dfrac{1.92}{159.6}$	$\dfrac{1.08}{18}$
	$= 0.012$	$= 0.06$

Step 2 Find the ratio of the moles of each substance.

	$CuSO_4$	H_2O
moles	0.012	0.06
divide by smallest	$\dfrac{0.012}{0.012}$	$\dfrac{0.06}{0.012}$
	$= 1$	$= 5$

Formula is $CuSO_4.5H_2O$.

Worked example 2

11.89 g of hydrated cobalt chloride; $CoCl_2.xH_2O$ were heated in a crucible until no further change in mass was obtained. On cooling the solid was re-weighed. 6.40 g of solid remained. Determine the value of x in the formula $CoCl_2.xH_2O$.

Step 1 Write an equation.	$CoCl_2.xH_2O$	$\rightarrow$	$CoCl_2$	$+$	xH_2O
Step 2 Write data under the equation.	11.89 g		6.40 g		$11.89 - 6.40 = 5.49$ g
Step 3 Convert the data to moles.			0.049		0.305
Step 4 Determine the ratio.			1		6.2
Step 5 Round to the nearest whole number.			1		6
Step 6 Write out the formula.	$CoCl_2.6H_2O$				

Questions

1 Memorize the formulae of all the acids, bases, and alkalis referred to on this double page. You must be able to write all of these formulae confidently.

2 Write balanced equations for the following reactions.

  a sodium oxide with hydrochloric acid

  b magnesium hydroxide with sulfuric acid

  c ammonia solution with nitric acid

  d solid magnesium carbonate with hydrochloric acid

3 6.45 g of hydrated magnesium sulfate, $MgSO_4.xH_2O$, were heated until all the water of crystallization was driven off. 3.16 g of anhydrous magnesium sulfate remained. Determine the formula of the hydrated magnesium sulfate.

1.07 Redox reactions

Oxidation

Oxidation is the loss of electrons.

During oxidation the oxidation number of a species will increase.

Metals typically form positive ions during an oxidation reaction.

$Mg \rightarrow Mg^{2+} + 2e^-$

An **oxidizing agent** is a species that oxidizes another substance by removing electrons from it.

Reduction

Reduction is the gain of electrons.

During reduction the oxidation number of a species will decrease.

Non-metals typically react by forming negative ions during a reduction reaction.

$O_2 + 4e^- \rightarrow 2O^{2-}$

A **reducing agent** is a species that reduces another substance by adding electrons to it.

Names and oxidation states

Roman numerals are used to show the oxidation state of an element where the name does not make this clear.

For example:
Nitrate(III) contains nitrogen in a 3+ oxidation state, NO_2^-.

Nitrate(V) contains nitrogen in a 5+ oxidation state, NO_3^-.

Oxidation and reduction

Oxidation reactions involve the loss of electrons; reduction reactions involve the gain of electrons. Oxidation and reduction always occur together, and reactions of this type are called redox reactions.

Oxidizing agents bring about oxidation. Common oxidizing agents include oxygen, chlorine, potassium dichromate(VI), and potassium manganate(VII). Reducing agents bring about reduction. Common reducing agents include group 1 and 2 metals, hydrogen, carbon, and carbon monoxide.

Oxidation number rules

There are several **oxidation number** rules that you must be able to apply. Work through each rule and the example carefully.

- Any pure element has an oxidation number of 0, e.g. nitrogen in N_2 is 0.
- Any monoatomic ion has an oxidation number equal to the charge on the ion, e.g. magnesium in Mg^{2+} has an oxidation number of 2+.
- In compounds, group 1 and group 2 ions have an oxidation number equal to the group number of the ion, e.g. sodium in NaCl has an oxidation number of 1+.
- In compounds, the most electronegative element fluorine has an oxidation state of 1−, e.g. fluorine in HF has an oxidation number of 1−.
- For all the compounds you will meet in AS, hydrogen has an oxidation state of 1+, e.g. hydrogen in H_2O has an oxidation number of 1+.
- In compounds the oxidation state of oxygen is 2− unless it is with fluorine, e.g. oxygen in H_2O has an oxidation number of 2−; oxygen in F_2O has an oxidation number of 2+.
- For neutral molecules, the sum of the oxidation numbers of the atoms is 0.
- For molecular ions, the sum of the oxidation numbers of the atoms is equal to the overall charge on the ion, e.g. for the CO_3^{2-} ion the sum of the oxidation numbers is 2−.

Examples of applying oxidation states

Sodium nitrate

Sodium nitrate has the formula $NaNO_3$.

- Sodium nitrate has no overall charge so the sum of the oxidation numbers of the atoms must be equal to zero.
- There are three oxygen atoms which each have a 2− charge.
- Sodium is in group 1 of the period table so it has a 1+ charge.
  - Ⓡ As a result the oxidation state of nitrogen must be 5+.

Carbonate

A carbonate ion has the formula CO_3^{2-}.

- Carbonate has an overall charge of 2− so the sum of the oxidation numbers of the atoms must be equal to 2−.
- There are three oxygen atoms which each have a 2− charge.
  - Ⓡ As a result the oxidation state of carbon must be 4+.

Writing formulae from oxidation numbers

For molecular ions in which the central atom can have a variety of oxidation numbers the oxidation number is given in brackets using a roman numeral. This information can be used to determine the formula of the molecular ion. Work through the example showing how to determine the formula of a molecular ion from a name.

What is the formula of the chlorate(V) ion?

- Ions with a name ending in -ate are always negative ions containing oxygen.
- Chlorate ions contain chlorine and oxygen.
- The chlorine has an oxidation number of 5+.
- Oxygen in compounds with chlorine has an oxidation number of 2–.
- To obtain a chlorine with an oxidation number of 5+ we need to use three oxygen atoms ($3 \times 2- = 6-$) and place a charge of 1– on the ion overall.
- The formula of the chlorate(V) ion is therefore ClO_3^-.

Identifying oxidation and reduction reactions

When magnesium is added to sulfuric acid a neutralization reaction occurs releasing hydrogen gas.

The overall equation for the reaction is

$$Mg(s) + H_2SO_4(aq) \rightarrow MgSO_4(aq) + H_2(g)$$

- The oxidation number of uncombined magnesium is 0; in $MgSO_4$ it is 2+.
- The oxidation number has gone up, so the magnesium is oxidized.
- The oxidation number of hydrogen in sulfuric acid is 1+; in hydrogen gas it is 0.
- The oxidation number has gone down, so the hydrogen is reduced.
- The electron transfer can be shown by splitting the overall equation into two **half equations:**

$$Mg \rightarrow Mg^{2+} + 2e^- \quad \text{(oxidation)}$$
$$2H^+ + 2e^- \rightarrow H_2 \quad \text{(reduction)}$$

The magnesium has reduced the hydrogen ions so it is a reducing agent. The hydrogen ion has oxidized the magnesium so is an oxidizing agent.

Work through this example showing how you can identify which species are being oxidized and reduced in a reaction through the use of oxidation numbers.

The reaction of iodide ions with chlorine gas shows an observable reaction forming iodine with chloride ions.

The overall equation for the reaction is:

$$Cl_2 + 2I^- \rightarrow 2Cl^- + I_2$$

- The chlorine has changed oxidation number from 0 to 1–.
- The chlorine has been reduced.
- Chlorine is acting as the oxidizing agent.
- The iodide has changed oxidation number from 1– to 0.
- The iodide has been oxidized.
- Iodide is acting as the reducing agent.

Worked example

State the oxidation number of sulfur in

a H_2SO_4

b H_2S

c SO_2

Answer

a
$$H_2SO_4 \quad \text{So S must be 6+}$$
$$1+ \times 2 \qquad 2- \times 4$$
$$= 2+ \qquad = 8-$$

b
$$H_2S \quad \text{So S must be 2–}$$
$$1+ \times 2$$
$$= 2+$$

c
$$SO_2 \quad \text{So S must be 4+}$$
$$2- \times 2$$
$$= 4-$$

State the oxidation number of iodine in

a KIO_3

b KI

c KIO_4

Answer

a 5+

b 1–

c 7+

Questions

1 Determine the oxidation number of the named element in each substance.

  a manganese in MnO_2

  b fluorine in F_2O

  c lithium in LiCl

  d calcium in $Ca(OH)_2$

  e sulfur in SO_4^{2-}

2 Write down the formula of:

  a manganate(VII)

  b sulfate(IV)

3 Bromide ions undergo a displacement reaction with chlorine molecules.

  a Write a balanced equation for this reaction.

  b Using oxidation numbers identify which species have been oxidized and which are reduced in this reaction.

  c Write down the name of the oxidizing agent and the reducing agent in this reaction.

1.08 Ionic bonding

Ionic bond

An electrostatic attraction between ions of opposite charge.

Formulae of some common ions

For most ions you can work out the charge using the periodic table.

- Metal ion: charge on the ion = group number of atom.
- Non-metal ion: charge on the ion = group number of atom − 8.
- For elements in the d-block you can work out the charge on the ion from the roman numeral given in brackets.

Practise this using the table below.

You will need to memorize the formulae for the molecular ions shown in bold on the table below.

Positive ions (cations)

hydrogen	H^+
sodium	Na^+
silver	Ag^+
potassium	K^+
lithium	Li^+
ammonium	$\mathbf{NH_4^+}$
barium	Ba^{2+}
calcium	Ca^{2+}
copper(II)	Cu^{2+}
magnesium	Mg^{2+}
zinc	Zn^{2+}
lead	Pb^{2+}
iron(II)	Fe^{2+}
iron(III)	Fe^{3+}
aluminium	Al^{3+}

Negative ions (anions)

chloride	Cl^-
bromide	Br^-
fluoride	F^-
iodide	I^-
hydroxide	$\mathbf{OH^-}$
nitrate	$\mathbf{NO_3^-}$
oxide	O^{2-}
sulfide	S^{2-}
sulfate	$\mathbf{SO_4^{2-}}$
carbonate	$\mathbf{CO_3^{2-}}$

Formation of ions

All chemical bonds are forces of attraction.
Ionic bonds occur when a metal and a non-metal react to form a compound.

- Metal atoms lose electrons forming **cations**.
- For example a lithium atom loses 1 electron to form a lithium ion.
  $Li \rightarrow Li^+ + e^-$ The lithium ion has electron configuration $1s^2$.
- This ion is **isoelectronic** (has the same electron configuration) with helium.
- Non-metal atoms gain electrons forming **anions**.
- For example a fluorine atom gains 1 electron to form a fluoride ion.
  $F + e^- \rightarrow F^-$ The fluoride ion has electron configuration $1s^2\ 2s^2\ 2p^6$.
- This ion is isoelectronic with neon.

Ions are formed by the transfer of electrons from the metal atom to the non-metal atom. This is shown in a dot and cross diagram:

- Only outer shell electrons are shown in a dot and cross diagram.
- Sodium chloride is formed by the transfer of an electron from sodium to chlorine.
- The electron which has come from the sodium is shown as a dot in the dot and cross diagram.
- There is no difference between an electron shown by a dot and an electron shown by a cross. They are just symbols used to show where electrons have come from.

Dot and cross diagrams

An ionic bond is the **electrostatic attraction** between cations and anions.
The ions in an ionic compound form a repeating three-dimensional structure, called a **lattice**.

- Dot and cross diagrams for an ionic bond simply show the cation and the anion.
- Think carefully about which group an atom is in to decide what the charge will be on the ion that is formed.
- Remember that when making ionic bonds atoms achieve full outer energy levels.

In these ionic compounds, different numbers of electrons are transferred depending on the group in which each atom is found.

Ionic lattices

Ionic compounds are formed of lattice structures.
- In an ionic compound each ion is surrounded by ions of opposite charge.
- For example, in the sodium chloride lattice, each chloride ion is surrounded by six sodium ions and vice versa.
- This structure is repeated throughout the ionic compound.
  - Ⓡ As a result ionic compounds are said to have giant structures.

Constructing ionic formulae

Ionic compounds are not molecules.
- You cannot write a molecular formula for an ionic compound.
- Instead you write an empirical formula.
- This shows you the ratio of cations to anions present in the ionic lattice.

A formula for an ionic compound is constructed from the formulae of the cations and anions present. In most cases you can work out the charges of the cation and anion using the periodic table.
- For metal atoms the charge on the ion is the same as the group number of the metal.
- For non-metal ions the charge on the ion is the same as the group number of the atom minus 8.
- Ionic compounds have a neutral charge overall so the number of positive charges must match the number of negative charges.

For example:
- Barium chloride contains Ba^{2+} ions and Cl^- ions, so has the formula $BaCl_2$.
- Lithium oxide contains Li^+ ions and O^{2-} ions, so has the formula Li_2O.

Some ionic compounds contain compound ions or molecular ions. It is worth learning the most common of these.

For example:
- Sodium hydroxide contains Na^+ ions and OH^- ions, so has the formula $NaOH$.
- Sodium sulfate contains Na^+ ions and SO_4^{2-} ions, so has the formula Na_2SO_4.
- Aluminium sulfate contains Al^{3+} ions and SO_4^{2-} ions, so has the formula $Al_2(SO_4)_3$.

Sodium chloride structure

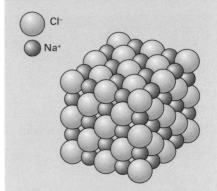

Cl^-

Na^+

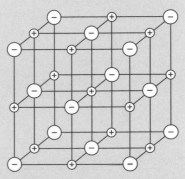

A space-filling diagram showing the lattice structure of sodium chloride.

This is the easiest way to draw the lattice structure of sodium chloride. Do practise drawing this.

Questions

1 Draw a dot and cross diagram to show the ionic bond in
  a sodium oxide
  b magnesium chloride
  c barium sulfide
2 Sketch the structure of sodium chloride until you can draw it from memory.
3 Construct formulae for
  a sodium chloride
  b magnesium bromide
  c strontium nitrate
  d calcium hydroxide
  e barium carbonate

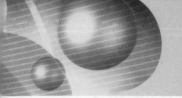

1.09 Covalent bonding

Covalent bonds

Single covalent bond: A shared pair of electrons.

Double covalent bond: Two shared pairs of electrons.

Triple covalent bond: Three shared pairs of electrons.

Sharing of electrons

Non-metal atoms can achieve full outer shells either by accepting electrons from metal atoms or by sharing pairs of electrons.

- **Covalent bonds** can form between identical atoms or different atoms.
- As with ionic bonding only the outer energy level electrons are involved in bonding.
- A pair of electrons is shared between two atoms.
- The bond is held together by the attraction between each nucleus involved in the bond and the pair of electrons.

Covalent bonds can be represented by dot and cross diagrams. They can also be shown as a straight line between the atoms. This is called a displayed formula.

Multiple bonds

Non-metal atoms can form double or triple covalent bonds by sharing more than one pair of electrons. These are represented in molecular formulae by multiple lines between the atoms. Double and triple bonds are stronger than single bonds.

Dative covalent bonds

Lone pairs

Pairs of electrons that are not involved in bonding are called **lone pairs** of electrons.

- Lone pairs of electrons are important in determining the shapes of molecules.
- Lone pairs also influence some chemical properties of molecules.
- Ammonia has one lone pair of electrons since nitrogen is in group 5 and forms single covalent bonds with three hydrogen atoms.
- Water has two lone pairs since oxygen is in group 6 and forms single covalent bonds with two hydrogen atoms.

Bonding with lone pairs

Lone pairs of electrons are able to form **dative covalent** (or co-ordinate) bonds with atoms that have vacant orbitals.

- A dative covalent bond is a shared pair of electrons in which both electrons are contributed by one of the atoms in the bond.
- Dative covalent bonds are shown in displayed formulae by an arrow.

For example:

- The ammonium ion, NH_4^+, has a dative covalent bond between the nitrogen atom and one of the hydrogen atoms.
- In carbon monoxide, CO, the oxygen atom forms a double covalent bond and also a dative covalent bond.
- In the NH_3BH_3 molecule the boron atom has a vacant orbital and accepts a pair of electrons from the nitrogen atom.

Giant covalent bonding

Molecules such as O_2, NH_3, CO_2, and H_2O are described as simple molecules. These are discrete molecules made from a small number of atoms.

- Carbon forms a number of structures called **allotropes**.
- These are different structures of the same element.
- Each carbon atom can form four covalent bonds.
- This ability to form a number of bonds enables carbon to bond to itself.
- Diamond and graphite are two different allotropes of carbon each of which has a giant molecular structure.
- A giant molecule is one in which the same arrangement of atoms is repeated many times.

Diagrams of diamond and graphite are on spread 1.15.

Dative covalent bonding

Lone pair of electrons: A pair of electrons not involved in bonding.

Dative covalent bond (or co-ordinate bond): A shared pair of electrons in which one of the atoms contributes both electrons.

Diamond

This is the hardest natural substance because of its structure and bonding.

Expanding their octet

Elements in period 3 of the periodic table are able to have more than eight electrons in their outer energy level. This is because they have access to the d sub-shell which can contain ten electrons.

You are most likely to come across sulfur as an example of an element which can do this. In sulfur dioxide, for example, the sulfur atom has ten electrons in its highest energy level. Note that oxygen cannot expand beyond eight electrons as it is a period 2 element.

Note also that elements in group 3 can have vacant orbitals in their compounds.

For example: in boron trifluoride, BF_3, boron has six electrons in its highest energy level.

Questions

1. Draw dot and cross diagrams to show the bonding in:
   a. Cl_2
   b. CCl_4
   c. PH_3
   d. H_2S
2. Draw dot and cross diagrams and displayed formulae to show the bonding in
   a. $NH_3 \cdot AlCl_3$
   b. $PH_3 \cdot BF_3$
3. Sketch a small portion of the structures of diamond and graphite. Add labels to each diagram to explain the properties of each allotrope.

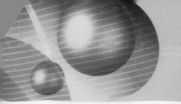

Molecular shapes

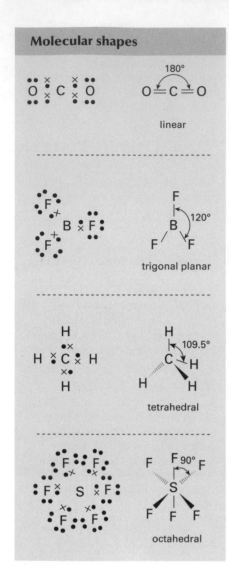

linear

trigonal planar

tetrahedral

octahedral

Valence shell electron pair repulsion theory

The shapes of molecules and ions are determined by applying valence shell electron pair repulsion theory:

- **Valence shell electrons** are those in the outermost energy level.
- These pairs of electrons repel each other.
- The shape of a molecule is such that the distance between the pairs of electrons is as large as possible.
- Multiple bonds have the same repelling effect as single bonds.
- **Lone pairs of electrons** are more repelling than bonding pairs of electrons.
  - ® As a result the bond angles are slightly smaller when lone pairs are present.

To work out the shape of a molecule you must first draw a dot and cross diagram. Remember that this diagram only includes electrons in the outermost energy level. From this diagram, you can determine the number of pairs of electrons around the central atom and hence the shape.

You must be able to remember the shapes of molecules shown in the table below.

pairs of electrons	basic shape	bond angles
2	linear	180°
3	trigonal planar	120°
4	tetrahedral	109.5°
6	octahedral	90°

For molecules that are not flat a standard notation is used for showing their shape:

- The dotted line shows a bond going behind the plane of the paper.
- The shaded triangle shows a bond coming out of the plane of the paper.
- The ordinary line shows a bond in the plane of the paper.

It is always advisable to sketch and name a shape in an examination. This helps if your drawing skills are not particularly strong!

Molecules containing one or more lone pairs of electrons

Tetrahedral

Both ammonia, NH_3, and water, H_2O, are molecules containing four pairs of electrons overall.

- In ammonia one of these pairs of electrons is a lone pair.
  - ® As a result ammonia has a pyramidal shape.
- In water two of the pairs of electrons are lone pairs.
  - ® As a result water has a non-linear shape.
- The repulsive effect of a lone pair of electrons is greater than a bond pair.
  - ® As a result the bond angles in ammonia and water are slightly smaller than that of methane.

pyramidal

non-linear

Trigonal planar

Sulfur dioxide is a molecule containing two double bonds and one lone pair of electrons.

- The shape of the sulfur dioxide is based on a trigonal planar shape.
- One of the corners of the molecule is a lone pair.
- As a result the shape of sulfur dioxide is V-shaped.
- The S=O bond angle is slightly less than 120°.

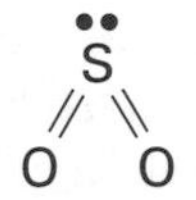

V-shaped

Shapes of ions

The shapes of molecular ions can be worked out in the same way as for uncharged molecules.

- Remember to take account of the charge on the ion when working out the dot and cross diagram.
- Positively charged ions have fewer electrons than the original atom.
- Negatively charged ions have gained extra electrons. These can be indicated on a dot and cross diagram using a triangle.

Examine the shapes of the ions carefully. Note the following points:

- In the NH_4^+ ion there is a dative covalent bond between the nitrogen and one of the hydrogen atoms.
- In the H_3O^+ ion there are three bonding pairs of electrons and one lone pair around the central oxygen atom.
- As a result the shape of the molecule is pyramidal with H–O bond angles of 107°.

The sulfate ion, SO_4^{2-} has a complex dot and cross diagram:

- The sulfur atom forms two double bonds and two single bonds.
- The two electrons from the two minus negative charge are placed on the singly bonded oxygen atom in order to give the oxygen a complete outer energy level.
- The electrons in the S=O bonds are delocalized (spread out) over the structure.
  - As a result the shape of the molecule is tetrahedral with equal bond angles of 109.5°.

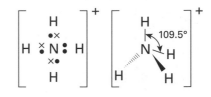

tetrahedral

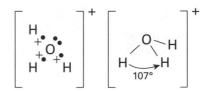

pyramidal

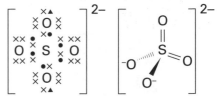

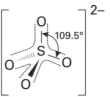

tetrahedral

Questions

1 Draw a dot and cross diagram for each of the following molecules. Then use it to draw a labelled diagram of the molecule showing all bond angles.

  a BF_3 b PF_5 c PH_3
  d H_2S

2 Draw a dot and cross diagram for the following ions. Remember to take the charges into account. Use the dot and cross diagram to draw each ion.

  a OH^- b PH_4^+ c PF_6^-

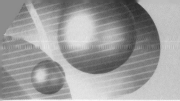

1.11 Electronegativity, polar bonds, and polar molecules

Charge distribution in covalent molecules

Covalent bonding is the sharing of one or more pairs of electrons.
The covalent bond is held together by the attraction between the nuclei of the two atoms involved in the bond and the pairs of electrons.

- If the bond is between identical atoms this sharing is equal e.g. H_2, Cl_2.
- If the atoms are different then the sharing may be unequal e.g. H–F.
- The fluorine atom is much better at attracting the pair of electrons than hydrogen because it has a greater number of protons.

Electronegativity

Electronegativity is the ability of an atom to withdraw electron density from a covalent bond.

- If the two atoms in a bond have different electronegativites then the more electronegative element has a greater share of the electrons.
- The electronegativity of an element can be calculated.
- The most common scale for electronegativity is the Pauling electronegativity scale.

You do not need to remember this scale but do need to understand that non-metal elements at the top of groups 5, 6, and 7 are the most electronegative.

Electronegativity trends

There are two main trends in electronegativity:

- Electronegativity increases across the periodic table as the number of protons in the nucleus increases. This increases the ability of an atom to attract electrons.
- Electronegativity decreases down the periodic table as the amount of electrons in complete energy levels increases. As a result the nucleus is shielded and has less ability to attract electrons.
- Fluorine is the most electronegative element owing to its small size.

Polar bonds

Polar bonds are those in which the pair of electrons is not shared equally.

- The more electronegative element has a partial negative charge, shown by δ–.
- The less electronegative element has a partial positive charge, shown by δ+.

In the hydrogen chloride molecule:

- The chlorine atom is more electronegative so has a partial negative charge.
- The hydrogen atom therefore has a partial positive charge.
  - As a result, The H–Cl *molecule* can be described as polar. $\overset{\delta+}{H} - \overset{\delta-}{Cl}$

In the carbon dioxide molecule:

- The oxygen atoms are more electronegative so have a partial negative charge.
- The carbon atom has two partial positive charges as it is bonded to two oxygen atoms.
  - As a result, The C=O *bond* can be described as polar. $\overset{\delta-}{O} = \overset{2\delta+}{C} = \overset{\delta-}{O}$

The Pauling scale for electronegativity							
H 2.2							He –
Li 1.0	Be 1.6	B 2.0	C 2.5	N 3.0	O 3.4	F 4.0	Ne –
Na 0.9	Mg 1.3	Al 1.6	Si 1.9	P 2.2	S 2.6	Cl 3.2	Ar –
K 0.8	Ca 1.0					Br 3.0	Kr 3.0
Rb 0.8						I 2.7	Xe 2.6

Polar bonds and polar molecules

A molecule that contains polar bonds may not be a polar molecule.

To find out if a molecule is polar:

- Draw the molecule (in three dimensions if necessary, remembering about the influence of lone pairs).
- Label any polar bonds using the $\delta+$, $\delta-$ convention.
- Then examine the shape of the molecule.
- If the polar bonds cancel out then the molecule is not polar.
- If the polar bonds do not cancel out then the molecule is polar.

Carbon dioxide has two polar bonds but there is no net dipole as the bonds are symmetrical. Carbon dioxide is a non-polar molecule.

Boron trifluoride has three polar bonds but there is no net dipole as the bonds are symmetrical. Boron trifluoride is a non-polar molecule.

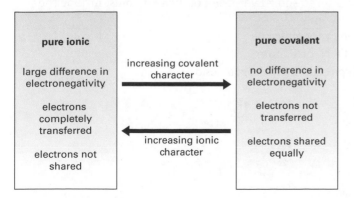

Water has two polar bonds and has a net dipole beacause the polar bonds do not cancel each other out. Water is polar molecule. This has a big influence on the properties of water.

The character of bonds

Covalent bonds

The greater the electronegativity difference between two atoms the bigger the **dipole** (difference in charge between the atoms) and the more polar the bond.

For example:

- The H–F bond has an electronegativity difference of 1.8.
- The H–Cl bond has an electronegativity difference of 1.0.
- The H–Br bond has an electronegativity difference of 0.8.
  - As a result the H–F bond is the most polar

If there is a large difference in electronegativity, the covalent bond can be described as having ionic character. So H–F has the most ionic character.

- If there is a very large difference in electronegativity then the bonding is ionic rather than covalent.
- It is a good idea to think of ionic and covalent bonding as the extreme ends of a range of types of bonding.

pure ionic		pure covalent
large difference in electronegativity	increasing covalent character →	no difference in electronegativity
electrons completely transferred		electrons not transferred
electrons not shared	← increasing ionic character	electrons shared equally

Ionic bonds

Ionic bonds can also show polarity. You will not be examined on this section but it does help explain the rather surprising bonding of some of the compounds.

- The electron cloud around a large negative ion can be distorted by a small highly charged positive ion (these have high charge density).
- The positive ion is said to be **polarizing.**
- The negative ion is **polarized.**
- If this happens to a large enough extent the ionic bond takes on covalent character.

For example:

- Sodium ions, Na^+, have a low charge density.
- This means they are not able to polarize chloride ions.
- Sodium chloride has ionic bonding.
- Aluminium ions, Al^{3+}, have a high charge density.
- This means they are able to polarize chloride ions.
- This polarization is so great that aluminium chloride has covalent bonding.

Questions

1 Label the following bonds to indicate their polarity.
   a H–F
   b C–Cl
   c O–N
   d S=O

2 Draw the following molecules. Label any polar bonds and state whether the molecule is polar.
   a hydrogen bromide
   b hydrogen sulfide
   c ammonia
   d fluorine oxide, F_2O

3 Define the term electronegativity and explain the trend in electronegativity down group 7.

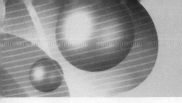

Intermolecular forces

An intermolecular force is a weak attractive force between molecules.

A permanent dipole–dipole attraction is an attractive force that exists between polar molecules.

A hydrogen bond is an intermolecular force between a lone pair of electrons on an N, O, or F atom in one molecule, and an H atom joined to an N, O, or F atom in another molecule.

Intermolecular forces are weak attractions that exist between molecules. These are the forces that hold substances together in the solid or liquid form. The strength of these forces determines the melting points and boiling points of substances and can influence some of their other properties too.

Permanent dipole–dipole forces

Molecules with a **permanent dipole** have regions of different electron density within them. These molecules are described as being polar.

It is easy to see that if two polar molecules come near each other in space there will be an attraction between them.

For the hydrogen chloride molecule below, the electronegative chlorine of one HCl molecule will attract the electropositive hydrogen of another.

$$\overset{\delta+}{H} — \overset{\delta-}{Cl} \text{---------} \overset{\delta+}{H} — \overset{\delta-}{Cl} \text{---------} \overset{\delta+}{H} — \overset{\delta-}{Cl}$$

permanent dipole-dipole force

- In liquid HCl these attractions are constantly breaking and re-forming as the molecules move around slowly.
- In solid HCl these attractions hold the molecules in a fixed position.
- This has a big influence on the boiling point of hydrogen chloride.

Forces of this type exist between any two molecules that have permanent dipoles.

Hydrogen bonding

Hydrogen bonds are an especially strong **permanent dipole–dipole force**, which exists between molecules that contain very electronegative elements.

For a hydrogen bond to occur:

- The molecule must have an O, N, or F atom bonded to a hydrogen atom.
- The molecule to which it is attracted must contain an O, N, or F atom.
- O, N, and F are the most electronegative atoms.
   - As a result the dipole–dipole force between the molecules is especially strong.

The most common examples of hydrogen bonding asked about in AS exams are

- water
- ammonia
- hydrogen fluoride

There are, of course, many other molecules that are capable of hydrogen bonding and you will meet some of these in A2 and in other science courses. For example, protein chains are linked together by hydrogen bonds.

When drawing a diagram to show hydrogen bonding you must include the following:

- labelled dipoles on every molecule
- lone pairs on the O, N, or F atom
- a dotted line to represent the hydrogen bond

Hydrogen bonding in water.

Hydrogen bonding and physical properties

Intermolecular forces have a big influence on physical properties. These are properties such as melting point, boiling point, density, etc.

The two boiling point graphs show the significantly larger boiling points for water and hydrogen fluoride in comparison to the hydrides of other members of the same group.

We will revisit these graphs in the next section to consider the rest of the pattern.

The boiling points of the group 6 hydrides

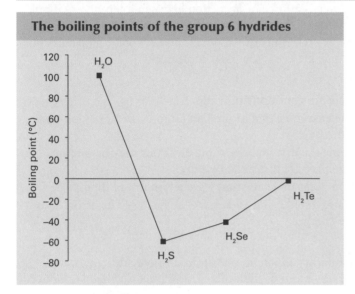

The boiling points of the group 7 hydrides

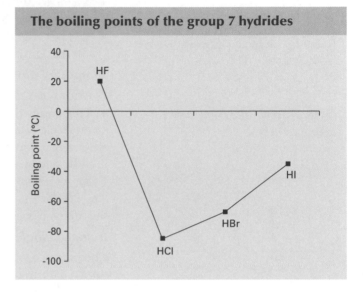

Ice

Ice has a very regular structure held together by hydrogen bonding between molecules. This hydrogen bonding causes water to have some unusual properties.

- Ice floats on water. It is the only substance for which this is the case.
- Ice is less dense than liquid water.
- Water has a large surface tension, caused by a network of hydrogen bonds on the surface.
- Water has a higher boiling point than would be expected. This is because hydrogen bonds must be broken when water is boiled.

The structure of ice

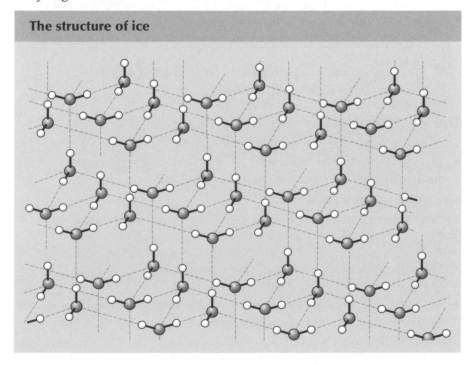

Questions

1 Draw a labelled diagram from memory to show the hydrogen bonding in water. Check your diagram against the book. Make sure that you have labelled the dipoles, shown lone pairs, and labelled your dotted line.

2 Draw a labelled diagram to show the hydrogen bonding in

   a hydrogen fluoride

   b ammonia

3 Make a list of the unusual properties of water. Write a one sentence explanation of each one.

Temporary dipole–induced dipole force

These forces are also known as van der Waals' forces

Temporary dipole: The asymmetrical distribution of the electron pair in a covalent bond.

Induced dipole: An uneven distribution of charge in a molecule or atom, caused by a charge in an adjacent particle.

The formation of van der Waals' forces

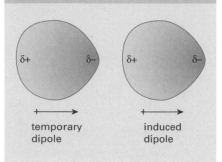

The structure of iodine

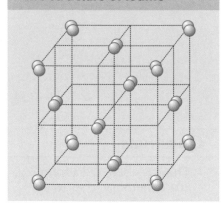

Van der Waals' forces

Molecules without permanent dipoles can form liquids and solids so there must be **intermolecular forces** between the molecules. Remember that it is intermolecular forces that are broken on melting or boiling, *not* covalent bonds.

*The name given to this force is a **van der Waals' force** after the Dutch scientist Johannes van der Waals who won the Nobel Prize in Physics in 1910.*

- A van der Waals' force is a force of attraction between a temporary dipole on one molecule and an induced dipole on another molecule.
- Van der Waals' forces do form between polar and non-polar molecules but are the dominant force between non-polar molecules.

Temporary dipoles

Electrons in a molecule are constantly moving.

- This means that the electron cloud around an atom or within a non-polar molecule is not static.
- At any instant in time the distribution of the electrons may be uneven although on average they are distributed evenly.
  - ℞ As a result a non-polar molecule may have a temporary dipole.

The dipoles can be represented using an arrow. The head of the arrow shows the region of negative charge.

Induced dipoles

The presence of a **temporary dipole** in one atom or molecule can cause a dipole to form in a nearby atom or molecule. This dipole is called an **induced dipole**. The induced dipole can then induce a dipole in a neighbouring atom or molecule. The net effect of this is a force of attraction between the particles called a **temporary dipole–induced dipole force**.

Van der Waals' forces and structure

All non-polar atoms or molecules have van der Waals' forces between the particles when they are in the liquid or solid state.

For example, iodine is a group 7 element that exists as diatomic molecules (two atoms joined together with a covalent bond).

- Iodine is a crystalline solid at room temperature.
- So the molecules in the iodine are in a regular arrangement, with van der Waals' forces between the molecules.
- These forces are relatively weak so little energy is needed to overcome them.
  - ℞ As a result iodine has a low melting point and boiling point.
- Iodine can sublime (change directly from a solid to a gas).

The strength of van der Waals' forces

The van der Waals' force strength is dependent on

- the size of the atom or molecule
- the area of contact between the atoms or molecules

Size of the atom or molecule

Look at the trend in boiling points of the group 7 elements shown in the graph. Work carefully through these points, which often comes up in AS examinations.

- The boiling points increase significantly down group 7.
- The size of the halogen molecules increases down group 7.
- The number of electrons in each molecule increases down the group.
  - ℞ As a result temporary dipoles form more readily in the large halogen molecules.

• In addition dipoles are more readily induced in adjacent molecules.
  🅑 As a result the van der Waals' forces get stronger down the group.
The same trend occurs in the alkanes as the number of carbon atoms in the chain increases.

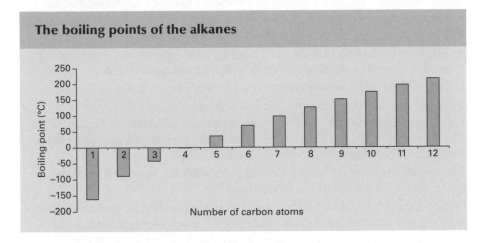

The boiling points of the alkanes

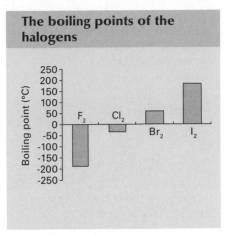

The boiling points of the halogens

Area of contact between molecules

The table of data below shows the boiling points of three alkanes. These all have the same molecular formula C_5H_{12} but have different chain lengths (you will revisit this work in the section on alkanes).

name	displayed formula	boiling point (°C)
pentane	H H H H H | | | | | H—C—C—C—C—C—H | | | | | H H H H H	36
2-methylbutane	H H H H | | | | H—C—C—C—C—H | | | | H H H H—C—H | H	28
2,2-dimethylpropane	H | H—C—H | H H | | H—C—C—C—H | | H H | H—C—H | H	10

The molecule with the lowest boiling point has the most chain branching. This increase in chain branching reduces the area of contact with other molecules and therefore reduces the strength of the van der Waals' force. Note – you do not need to remember this data but do need to remember the trend!

Questions

1 One of the first substances studied by Johannes van der Waals in the late 1800s was argon.

  a Complete the table below with your prediction for the melting point of argon.

  b Then write a one sentence explanation of your prediction.

element	atomic mass	melting point (°C)
helium	4	−272
neon	20	−249
argon	40	
krypton	84	−156
xenon	132	−112

2 Think about the boiling of bromine. Write a short paragraph explaining what happens to the molecules of bromine as it forms bromine gas.

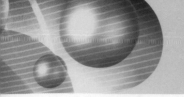

1.14 Metallic bonding

Metallic bond

Metallic bond: The electrostatic force of attraction between metal ions and the delocalized electrons in a metallic lattice.

Metallic bonding

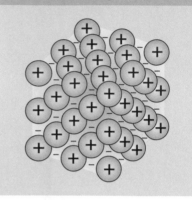

The metallic bond

Atoms of metals are held together by **metallic bonds**. In a metallic bond:
- Each metal atom forms a positive ion (cation).
- The positive ions are arranged into a regular **lattice** structure.
- The ions in the structure are very close to each other so the electrons that are lost when the metal atoms form ions become **delocalized**.
- Delocalized electrons are not attracted to any particular ion.
  Ⓡ As a result these electrons are free to move through the metal.

For example, magnesium atoms have the electron configuration $1s^2\ 2s^2\ 2p^6\ 3s^2$. The 3s electrons are delocalized from each atom forming magnesium ions of charge $2+$. These have the electron configuration $1s^2\ 2s^2\ 2p^6$.

The metallic bond is the strong attraction between the positive ions in the lattice and the delocalized sea of electrons.

Note – although the term lattice is used for ionic bonding as well as metallic bonding you must be careful not to confuse the two. The ionic lattice is a regular arrangement of ions of opposite charge. The metallic lattice is a regular arrangement of cations surrounded by a sea of delocalized electrons.

Metallic bond strength

Metallic bonds do not all have the same strength. If they did, then all metals would melt and boil at the same temperature.

The strength of a metallic bond is dependent on
- the charge of the ions in the lattice
- the number of electrons in the sea of delocalized electrons

The boiling point graph for sodium, magnesium, and aluminium shows this. Magnesium needs more energy to boil than sodium as the attraction between the Mg^{2+} ions in the lattice and the sea of electrons is greater than between the Na^+ and the sea. The force of attraction is even stronger in aluminium which has Al^{3+} ions.

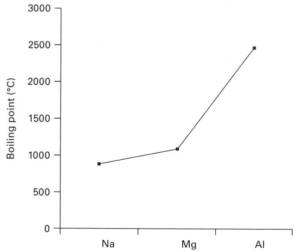

Boiling points of sodium, magnesium, and aluminium

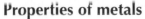

Properties of metals

The properties of metals can be easily described in terms of their lattice structure.

- Metals are **good conductors** of heat and electricity.
- Metals have high melting points and boiling points.
- Metals are **malleable**.
- Metals are **ductile**.

Conducting heat and electricity

Metals are good conductors of heat as the metal ions in the lattice are very close to each other. Heating one end of a piece of metal makes the ions at that end vibrate more, and these vibrations are passed along the piece of metal.

Metals are good electrical conductors as the sea of delocalized electrons is free to move when a voltage is applied. Remember that the electrons are not attracted to any particular ion. 'Pushing' electrons into one end of the piece of metal causes electrons to come out of the other end.

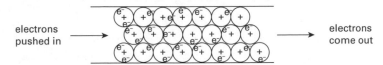

Malleability and ductility

Malleability is the ability of a piece of metal to be pressed or hammered into shape. Ductility is the ability to be stretched into a wire.

Both of these properties are due to the regular layer structure of the metal. The layers of cations can slip over each other if a large enough force is applied. The strength of the metallic bond stops the attractions being broken completely.

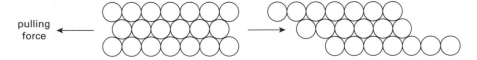

Questions

1 Draw a simple sketch of the structure of a piece of aluminium. Label your sketch to show how aluminium is able to conduct electricity.

2 List the following metals in order of their melting point. Explain the order you have chosen: sodium, aluminium, magnesium.

3 List the key properties of metals and write a two sentence explanation of each property in terms of the structure of the metal.

Sodium chloride structure

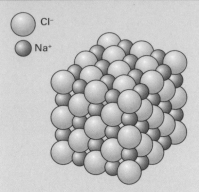

A space-filling diagram showing the lattice structure of sodium chloride.

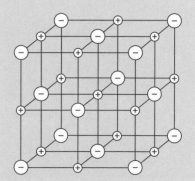

This is the easiest way to draw the lattice structure of sodium chloride. Do practise drawing this.

Metallic bonding

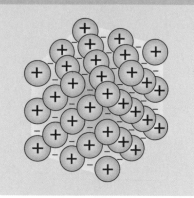

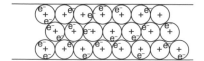

Changing state

When a solid is melted or a liquid frozen, a liquid boiled or a gas condensed it is said to have **changed state**.

- All changes of state involve changes in energy.
- When a solid melts or a liquid boils, the energy is used to break the forces between the atoms, molecules, or ions involved.
- As the change of state occurs, the temperature stays constant because the energy provided to the system is used to break the force.

It is essential that you are able to identify the type of force being broken in each case. Read this section after you have worked through the pages on ionic bonding, covalent bonding, metallic bonding, and intermolecular forces.

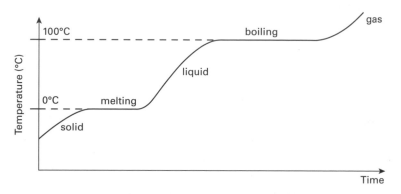

Metals

When melting a pure metal the attraction between the **lattice** of positive ions and the **delocalized sea** of electrons is broken. This is a strong force so requires a large amount of energy, which means that metals have high melting points. The melting points increase as the charge on the metal ion and the number of electrons in the sea increases.

Ionic substances

Remember that an ionic bond is an **electrostatic attraction** between ions of opposite charge. When an ionic substance melts, the energy provided is used to break this attraction. The attraction is strong so ionic substances are hard to melt.

Simple molecular substances

Melting simple molecular substances requires the breaking of the **intermolecular force** between the molecules.

- There are three types of forces that can be broken.
- Each of these forces is much weaker than the covalent bond that exists between the atoms within the molecule.

You must be able to identify the type of force between molecules and relate this to changes of state.

name of force	requirement for force	example of molecule	relative strength of force
permanent dipole–dipole attraction	polar molecule	HCl PH_3	stronger than van der Waals' force, weaker than hydrogen bond
hydrogen bond	O–H, F–H, N–H bond present in molecule	H_2O HF NH_3	the strongest intermolecular force
van der Waals' force	non-polar molecule	Cl_2, Br_2, I_2, CH_4, C_2H_6, etc.	weakest intermolecular force

Giant molecular substances

Giant molecular substances have a large network of covalent bonds. In order to melt these substances strong covalent bonds must be broken. This requires a large amount of energy.

You must be able to sketch diamond and graphite and be able to explain why they have such high melting points. Remember that silicon has the same structure.

In diamond:
- Each carbon atom forms four covalent bonds.
- The shape around each carbon atom is tetrahedral.
- The C–C bond angle about each atom is 109.5°.
- This tetrahedral structure is repeated around each carbon atom making diamond extremely strong.

In graphite:
- Each carbon atom forms three covalent bonds.
- The shape around each carbon atom is trigonal planar.
- The C–C bond angle about each atom is 120°.
- The carbon atoms form a flat lattice structure made from hexagons.
- The extra electron from each carbon is contributed to a delocalized sea of electrons between the layers.
- These electrons are free to move so graphite is able to conduct electricity.
- Weak van der Waals' forces exist between the layers.
- These weak forces allow the layers to slide over each other.
- As a result graphite is soft and slippery.

Conducting electricity

In order to conduct electricity charged particles must be able to move through a substance when a voltage is applied:
- In a metal the delocalized electrons are free to move.
- In a molten ionic substance the ions are free to move. Note that ionic solids do not conduct electricity as the ions are held in a fixed position.
- Graphite is able to conduct as there are delocalized electrons between the layers.

Solubility

You must be able to describe whether a substance will dissolve in water based on its bonding. Solubility is a

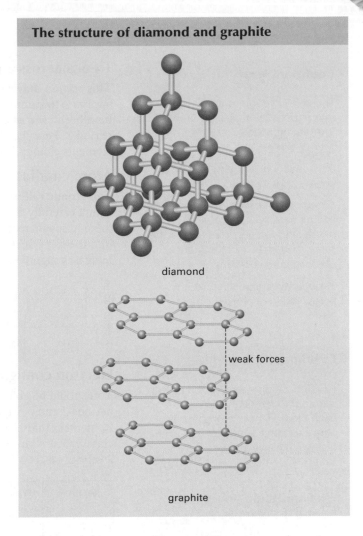

The structure of diamond and graphite

diamond

weak forces

graphite

complicated concept that you will learn more about later on in your studies.

Some simple rules to apply for AS chemistry are:
- Ionic substances are **soluble** (they dissolve) in water.
- Molecules that are very polar are soluble in water.
- Molecules with no polarity are **insoluble** (don't dissolve) in water.

For example, sodium chloride dissolves in water as it is ionic. Diamond does not dissolve in water as it has no polar bonds and has a giant covalent structure. Hydrogen chloride is soluble in water as it has very polar bonds.

Questions

1 Complete the table below showing the type of bonding in each substance.

name of substance	formula	type of structure	type of bonding
magnesium	Mg	giant	
sodium chloride			ionic
chlorine			
graphite			

2 Sketch a small section of diamond and of graphite. Use your diagrams to explain why graphite is able to conduct electricity and diamond is not.

3 Arrange the following substances in order of melting point starting with the lowest. Explain the order you have chosen. H_2O, $MgCl_2$, Cl_2.

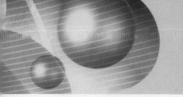

Important terms

The periodic table is arranged into periods by increasing atomic (proton) number.

Period: a row of elements showing repeating trends in physical and chemical properties.

Group: a column of elements having similar chemical and physical properties. Elements in the same group have similar outer shell electron configurations.

Periodicity: a repeating pattern across different periods.

First ionization energy

First ionization energy is the energy required to remove 1 electron from each atom in 1 mole of gaseous atoms forming 1 mole of ions with a single positive charge.

First ionization energy for period 3 elements

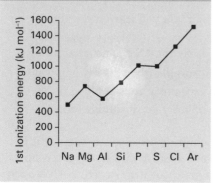

Trends across a period

This section draws together lots of the content on structure and bonding and applies it to examining trends across a period. The elements from period 3 have been used as examples; remember that the same trends occur across period 2. Periodic chemistry is very different from group chemistry as the elements change from metallic to non-metallic across a period.

Atomic radius

The **atomic radius** of the elements decreases across period 2 and period 3. Think carefully about this trend. Remember that the number of protons in the nucleus increases across the period and that the extra electrons are placed into the same shell. This means that the effect of the nucleus on the electrons increases slightly across the period, making the atomic radius slightly smaller.

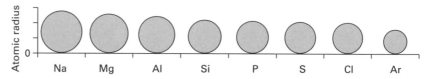

Electron configurations and first ionization energy

You should be able to write the electron configuration for all the atoms in periods 2 and 3. Practise reading this information from the periodic table. Remember that elements with outer electrons in the s energy level can be classified as s block elements and those with outer electrons in the p energy level can be classified as p block elements.

symbol	type of element	electron configuration	block in the periodic table
Na	metal	$1s^2\ 2s^2\ 2p^6\ 3s^1$	s block
Mg	metal	$1s^2\ 2s^2\ 2p^6\ 3s^2$	s block
Al	metal	$1s^2\ 2s^2\ 2p^6\ 3s^2\ 3p^1$	p block
Si	non-metal	$1s^2\ 2s^2\ 2p^6\ 3s^2\ 3p^2$	p block
S	non-metal	$1s^2\ 2s^2\ 2p^6\ 3s^2\ 3p^3$	p block
P	non-metal	$1s^2\ 2s^2\ 2p^6\ 3s^2\ 3p^4$	p block
Cl	non-metal	$1s^2\ 2s^2\ 2p^6\ 3s^2\ 3p^5$	p block
Ar	non-metal	$1s^2\ 2s^2\ 2p^6\ 3s^2\ 3p^6$	p block

It is essential that you can sketch and explain the graph showing the **first ionization energy** for the elements in period 2 and period 3. The key details are:

- There is a general increase across a period as the number of protons in the nucleus steadily increases, leading to a greater charge on the nucleus.
  - Ⓡ As a result there is a greater attraction between the nucleus and the outer electron.

You will not be examined on the explanation for the small decreases between groups 2 and 3 or 5 and 6.

First ionization energy decreases down a group:

- The atomic radius increases.
- Electron shielding increases.
- The number of protons in the nucleus increases.
- The increase in radius and electron shielding outweigh the increased nuclear charge.

Melting points and boiling points

Trends in melting point and boiling point across period 2 and period 3 are related to the change in bonding from metallic to covalent as you move across the period. It is important that you can

- state the type of bonding in each element
- sketch the structure of each element
- sketch and explain the melting point and boiling point graph.

element	bonding/structure	force broken on melting
sodium	metallic	metallic bond
magnesium	metallic	metallic bond
aluminium	metallic	metallic bond
silicon	giant covalent	covalent bond
phosphorus	simple molecular, P_4 units	van der Waals' force
sulfur	simple molecular, S_8 units	van der Waals' force
chlorine	simple molecular, Cl_2 units	van der Waals' force
argon	monatomic	van der Waals' force

Boiling points of the period 3 elements

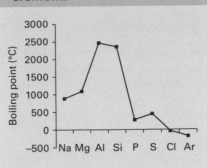

Structures of some period 3 elements

Phosphorus, sulfur, and chlorine are all simple covalent molecules with **van der Waals' forces** between them. This force strength increases with the increasing size of the molecule so the melting or boiling point increases in the order argon, chlorine, phosphorus, and sulfur.

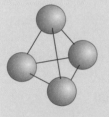

phosphorus

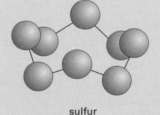

sulfur

chlorine

The bonding in silicon is **giant covalent** with a network of covalent bonds throughout the structure. Melting or boiling silicon requires the breaking of covalent bonds so requires a lot of energy. It is worth practising drawing a small section of silicon using shaded and dotted lines to show the tetrahedral structure about each silicon atom (which is exactly the same as diamond).

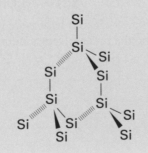

Exam style questions

Questions about melting point or boiling point always require you to identify the type of force being broken on changing state and to explain the strength of the force. Work through the sample question below.

Sodium melts at 98 °C, chlorine melts at −102 °C. Explain this difference in terms of structure and bonding.

Sodium has metallic bonds; chlorine has van der Waals' forces between molecules. Metallic bonds are a lot stronger than van der Waals' forces so much more energy is needed to break them. This gives sodium a significantly higher melting point than chlorine.

Questions

1. Practise sketching the curve for first ionization energy for period 3 elements. Cover up small sections of it then try and fill them in.

2. Draw an electron spin diagram for magnesium and aluminium. Use the diagram to explain the decrease in first ionization energy between magnesium and aluminium.

3. Draw the structures of silicon, sulfur, and phosphorus. Use them to describe the differences in boiling point.

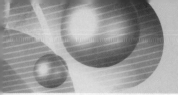

1.17 Group 2: the alkaline earth metals

Metallic bonding

- In metals the electrons in the outer shell of atoms are **delocalized**. This leads to positive metal ions and negatively charged delocalized electrons.

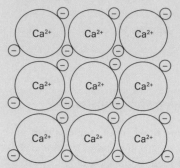

The metallic bonding in calcium metal.

- Metallic bonding is the electrostatic attraction between the positive ions and the negative delocalized electrons.
- Down the group the size of the metal ions increases so the strength of the metallic bonding decreases.
- Less energy is required to overcome the forces of attraction.
  - As a result melting points generally decrease down the group.

Atomic radius

As you go down group 2 the atoms have an extra shell of electrons and become larger.

As a result the **atomic radii** of the elements increase down group 2.

Ionization energy

Down the group the outer electrons are lost more easily:

- The number of protons increases down the group.
- The atomic radii increase down the group.
- This means the distance between the nucleus and the outer electrons increases.
- The amount of shielding by inner electrons increases.
- As a result the value for **first ionization energy decreases** down the group.

The alkaline earth metals

The elements in group 2 of the periodic table are called the alkaline earth metals.

- The group includes magnesium, calcium, strontium, and barium.
- The group 2 elements have metallic bonding.
- All of the group 2 elements have an outer shell electron arrangement of s^2.
- As a result they react to form ions that have a 2+ charge.

Reaction with oxygen

The group 2 elements react with oxygen to form oxides containing the group 2 element in a 2+ oxidation state and the oxide ion, O^{2-}. In these reactions the group 2 element is oxidized.

For example:

$$2Mg(s) + O_2(g) \rightarrow 2MgO(s)$$

The reactivity of the group 2 elements is dominated by the formation of the 2+ ion.

- The ionization energy decreases down the group.
- As a result the reactivity of the group 2 elements increases down the group.

Reaction with water

Group 2 metal atoms are good reducing agents. Their reaction with water is an example of a **redox** reaction.

- The group 2 metal atom is oxidized.

$$M \rightarrow M^{2+} + 2e^-$$

- The hydrogen in water is reduced.

Magnesium reacts slowly with water to form magnesium hydroxide and hydrogen.

$$Mg(s) + 2H_2O(l) \rightarrow Mg(OH)_2(aq) + H_2(g)$$

Magnesium hydroxide is sparingly soluble.
Magnesium reacts quickly with steam.

$$Mg(s) + H_2O(g) \rightarrow MgO(s) + H_2(g)$$

Calcium, strontium and barium all react vigorously with water to form a metal hydroxide and hydrogen.

Example

$$Ca(s) + 2H_2O(l) \rightarrow Ca(OH)_2(aq) + H_2(g)$$

A solution of calcium hydroxide looks cloudy as it is only slightly soluble in water and the undissolved calcium hydroxide forms a white suspension.

Reactions of group 2 oxides

The group 2 oxides react with water forming alkaline solutions containing group 2 hydroxides.

- Magnesium oxide reacts with water forming the slightly soluble magnesium hydroxide.

$$MgO(s) + H_2O(l) \rightarrow Mg(OH)_2(aq)$$

- Calcium oxide reacts with water forming an aqueous solution of calcium hydroxide (limewater).

$$CaO(s) + H_2O(l) \rightarrow Ca(OH)_2(aq)$$

The solution of calcium hydroxide has a pH between 9 and 11.

- If water is dropped slowly onto the oxide, slaked lime (solid calcium hydroxide) is formed.

$$CaO(s) + H_2O(l) \rightarrow Ca(OH)_2(s)$$

- Barium oxide and strontium oxide readily form the hydroxide with water.
- The group 2 hydroxides become more basic and more soluble down the group so the pH of the solution formed increases down the group.

Thermal decomposition of group 2 carbonates

The carbonates of group 2 elements decompose on heating forming the group 2 oxide and carbon dioxide gas.

For example:

$$CaCO_3(s) \rightarrow CaO(s) + CO_2(g)$$

- As you go down group 2 the carbonates become more stable.
- As a result the carbonates have to be heated more strongly in order to make them decompose.
- Note that this is not a redox reaction as none of the elements changes its oxidation state during the reaction.

Questions

1 Barium reacts much more readily with water than does magnesium. Explain this statement fully.

2 Write balanced equations for the following reactions:

  a magnesium and oxygen

  b strontium and water

  c the thermal decomposition of barium carbonate

3 Draw a dot and cross diagram to show the bonding in strontium oxide. Show outer shell electrons only.

4 Arrange the following elements in order of increasing ionization energy:

  calcium, strontium, magnesium, barium

Uses of group 2 hydroxides

Magnesium hydroxide is used in medicine to treat indigestion and heartburn.

- It is dissolved in solution to make milk of magnesia.
- Indigestion is caused by the stomach producing too much hydrochloric acid.
- Heartburn is caused when stomach acid is pushed into the oesophagus.
- The milk of magnesia neutralizes the excess acid:

$$Mg(OH)_2(s) + 2HCl(aq) \rightarrow MgCl_2(aq) + 2H_2O(l)$$

Calcium hydroxide is used in agriculture to neutralize acid soils.

- The calcium hydroxide can be formed by heating calcium carbonate to form calcium oxide and carbon dioxide.

$$CaCO_3(s) \rightarrow CaO(s) + CO_2(g)$$

- The calcium oxide then reacts with water to form calcium hydroxide, $Ca(OH)_2$.

$$CaO(s) + H_2O(l) \rightarrow Ca(OH)_2(s)$$

- Acids from acid rain and some fertilizers lower the pH of soil.
- Farmers lime soil using powdered limestone, calcium oxide, or calcium hydroxide to increase the pH of the soil.

Trends in thermal decomposition

Group 2 carbonates become more stable down group 2.

More heat energy is required to break up the carbonate for carbonates of metals further down group 2.

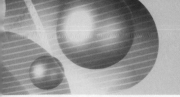

Introducing the halogens

- The elements in group 7 of the periodic table are often called the **halogens**.

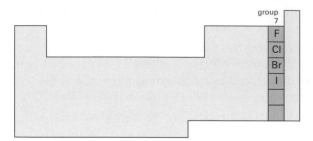

Group 7 elements are typical non-metals.

- Halogen atoms have seven electrons in their outer shell.

Example

When halogen atoms react they can gain an electron to form a halide ion which has a 1− charge.

Trends in electronegativity

Electronegativity is a way of measuring the attraction that a bonded atom has for the electrons in a covalent bond. Down the group the halogens become less electronegative.

Trends in boiling point

element	boiling point (°C)
fluorine	−188
chlorine	−35
bromine	+59
iodine	+216

At room temperature:
- Fluorine is a yellow gas.
- Chlorine is a pale green gas.
- Bromine is a brown liquid.
- Iodine is a dark grey solid.

Notice how the boiling points of the halogens increase down the group.

- The halogens exist as diatomic molecules, e.g. chlorine exists as chlorine molecules, Cl_2.
- This means that there are strong covalent bonds within the halogen molecules but only very much weaker van der Waals' forces of attraction between halogen molecules.
- The strength of the van der Waals' forces depends on the number of electrons in the molecules.
- Down the group the number of electrons in the halogen molecules increases.

  As a result down the group the boiling point increases.

Worked example

Complete the table below

element	state at room temperature	colour
fluorine		
chlorine		
bromine		
iodine		

element	state at room temperature	colour
fluorine	gas	yellow
chlorine	gas	pale green
bromine	liquid	brown
iodine	solid	dark grey

Worked example

Explain why iodine has a higher boiling point than fluorine.

Answer

Down the group the boiling point of the elements increases because there are stronger intermolecular forces between the molecules. Weak van der Waals' forces exist between the halogen molecules. Down the group there are more electrons in the atoms so there are stronger van der Waals' forces.

More energy is required to overcome these forces of attraction so the halogen boils at a higher temperature.

Trends in the oxidizing ability of the halogens

The **oxidizing ability** of a substance is a measure of the strength of an atom to attract and gain an electron.

- The halogens are good oxidizing agents.
- An oxidizing agent is a species which oxidizes another substance by removing electrons from it.
- Oxidizing agents are themselves reduced during the reaction.

When halogen atoms react they gain electrons.

Example

$$Cl_2 + 2e^- \rightarrow 2Cl^-$$

The oxidizing ability of the halogens decreases down the group.

Down the group the electron which is gained is being placed into a shell which is further from the nucleus.

- Down the group the atomic radii increases.
- The amount of shielding increases.

  - Ⓡ As a result the attraction between the nucleus and the electron decreases.
  - Ⓡ As a result the oxidizing ability of the halogens decreases down the group.

Displacement reactions

- The **displacement** reactions between halogens and aqueous halides demonstrate the decrease in the oxidizing ability of the halogens down the group.
- Chlorine is a more powerful oxidizing agent than bromine so chlorine oxidizes bromide ions.

Example

$$chlorine + bromide \rightarrow chloride + bromine$$

$$Cl_2(aq) + 2Br^-(aq) \rightarrow 2Cl^-(aq) + Br_2(aq)$$

The **half-equations** for this reaction are

$$Cl_2(aq) + 2e^- \rightarrow 2Cl^-(aq) \text{ (reduction)}$$

Chlorine is reduced to chloride.

$$2Br^-(aq) \rightarrow + Br_2(aq) + 2e^- \text{(oxidation)}$$

Bromide is oxidized to bromine.

Bromine is a more powerful oxidizing agent than iodine so bromine oxidizes iodide ions.

Example

$$bromine + iodide \rightarrow bromide + iodine$$

$$Br_2(aq) + 2I^-(aq) \rightarrow 2Br^-(aq) + I_2(aq)$$

The half-equations for this reaction are

$$Br_2(aq) + 2e^- \rightarrow 2Br^-(aq) \text{ (reduction)}$$

Bromine is reduced to bromide.

$$2I^-(aq) \rightarrow I_2(aq) + 2e^- \text{(oxidation)}$$

Iodide is oxidized to iodine.

Observations for displacement reactions

The formation of halogens in displacement reactions can be seen more clearly by adding a small volume of **cyclohexane** to the reaction mixture then shaking carefully.

The halogen molecule always dissolves in the cyclohexane; this layer floats on top of the aqueous layer, which always contains the halide ion.

For example:

- Cyclohexane is added to the reaction mixture from the reaction of bromine and iodide:
  $$bromine + iodide \rightarrow bromide + iodine$$
- The cyclohexane dissolves the iodine molecules forming a purple layer.
- The bromide ion remains in the aqueous solution.

You need to know the colours of the halogen molecules in cyclohexane.

halogen	colour of solution in cyclohexane
chlorine	pale green
bromine	orange
iodine	purple

Trends in reducing ability of halide ions

- The halogens are good oxidizing agents.
- The oxidizing ability of the halogens decreases down the group.
- The halide ions can act as reducing agents.
- A reducing agent is a species which reduces another substance by giving it electrons.
- Reducing agents are themselves oxidized during the reaction.
- Down the group the halide ions become increasingly good reducing agents.
- As a result iodide ions are the strongest reducing agents; iodide ions are easiest to oxidize.

Questions

1. Why does the boiling point of halogens increase down the group?
2. Explain what 'oxidizing ability' means
3. Why is chlorine a good oxidizing agent?
4. Write an ionic equation for the reaction of chlorine and iodide ions. Explain what you would see when cyclohexane is added to this reaction mixture.

1.19 Group 7: the halogens 2

Key terms

Work through the different forms of chlorine below. Make sure you use the correct one when answering questions. You will lose easy marks if you muddle them up. The same terms are used for all of the group 7 elements.

Cl   atom of chlorine

Cl_2 molecule of chlorine (this is how chlorine occurs naturally)

Cl^- chloride ion

ClO_3^- chlorate ion

Chlorate ions

Chlorate ions are negatively charged ions containing chlorine and oxygen.

Chlorate(I), ClO^-, contains chlorine in oxidation state 1+.

Chlorate(V), ClO_3^-, contains chlorine in oxidation state 5+.

Both of the chlorates are good oxidizing agents as they contain chlorine in a high oxidation state.

Identifying halide ions

We can identify halide ions using **acidified silver nitrate solution**.

- Fluoride ions form silver fluoride which is a soluble salt so no precipitate is seen.
- The other halide ions form insoluble salts. The colour of the silver salt can be used to identify the halide.

halide ion	salt formed	colour of the precipitate
chloride	AgCl	white
bromide	AgBr	cream
iodide	AgI	yellow

- The solubility of the silver halides in ammonia can be used to confirm the identity of the halide ions.

silver halide	solubility in ammonia
AgCl	Dissolves in dilute ammonia solution.
AgBr	Does not dissolve in dilute ammonia solution but does dissolve in concentrated ammonia solution.
AgI	Does not dissolve even in concentrated ammonia solution.

For example:

Sodium chloride solution reacting with silver nitrate solution

sodium chloride + silver nitrate → silver chloride + sodium nitrate

$$NaCl(aq) + AgNO_3(aq) \rightarrow AgCl(s) + AgNO_3(aq)$$

An ionic equation can be written for this reaction showing only the ions which undergo a change in the reaction:

$$Cl^-(aq) + Ag^+(aq) \rightarrow AgCl(s)$$

You do not need to be able to write equations for the precipitates dissolving in ammonia.

Worked example

A student is unsure if a solution is potassium bromide or potassium chloride. Describe how they could confirm the identity of the solution. Include any observations they would make and write a suitable ionic equation.

Answer

The student should add silver nitrate solution to the unknown solution. A white or cream precipitate will be seen. They should then add dilute ammonia solution. If the precipitate dissolves leaving a colourless solution then the unknown solution is silver chloride. If the precipitate does not dissolve fully then it is silver bromide.

Equation for the precipitate formation:

$$Ag^+(aq) + Cl^-(aq) \rightarrow AgCl(s)$$

Chlorine and water treatments

How does chlorine react with water?

Chlorine reacts with water to form two acids:

- chloric(I) acid
- hydrogen chloride.

$$Cl_2(aq) + H_2O(l) \rightleftharpoons HClO(aq) + HCl(aq)$$

The oxidation state of chlorine in

- Cl_2 is 0
- HClO is 1+
- HCl is 1−

Notice that this a disproportionation reaction. The chlorine is simultaneously oxidized (from 0 to 1+) and reduced (from 0 to 1−).

High levels of micro-organisms in unchlorinated water can cause devastating outbreaks of diseases such as cholera.

Sodium chlorate(I) is also used.

Chloric(I) acid ionizes to form hydrogen ions and chlorate(I) ions.

$$HClO(aq) \rightleftharpoons H^+(aq) + ClO^-(aq)$$

Chloric(I) acid is is much better at killing micro-organisms than chlorate(I) ions so the pH of the water should be kept lower than 8 to ensure that there is a high concentration of chloric(I) acid which makes the chlorination of water more efficient and the water safer to drink.

Problems with using chlorine

There are some concerns about the use of chlorine in water treatments.

- Chlorine is a toxic gas.
- Chlorine also reacts with organic compounds to form substances called chlorinated hydrocarbons which may be hazardous to humans although there is insufficient evidence to prove this link.

However, we continue to use chlorine because the health benefits of using chlorine far outweigh any possible problems it may cause.

The reaction of chlorine with sodium hydroxide

Chlorine reacts with cold, dilute sodium hydroxide to form sodium chloride, sodium chlorate(I), and water.

This is an example of **disproportionation**. The chlorine is simultaneously oxidized and reduced.

Sodium chlorate(I) is used to make household **bleaches**.

Worked example

The reaction between chlorine and water can be summed up by the equation.

$$Cl_2(aq) + H_2O(l) \rightleftharpoons HClO(aq) + HCl(aq)$$

Give the oxidation number of chlorine in:

a Cl_2 **b** HClO   **c** HCl

d Explain why the reaction between chlorine and water can be described as a disproportionation reaction.

Answer

a Cl_2 an elements has an oxidation number of 0

b   HClO   Cl must have an oxidation number of 1+

1+ ×1   2− ×1
= 1+   = 2−

c   HCl   Cl must have an oxidation number of 1−

1+ ×1
= 1+

d Because the chlorine is simultaneously oxidized (from 0 to 1+) and reduced (from 0 to 1−).

Chlorine in the UK

In the UK very low levels of chlorine are added to drinking water to reduce levels of micro-organisms to acceptable levels.

Questions

1. What is the trend in the strength of the ability of halide ions to reduce substances?
2. Using the information on this page write two equations each showing a disproportionation reaction of chlorine. Use oxidation numbers to show the chlorine has undergone disproportionation.
3. How could you identify bromide ions?

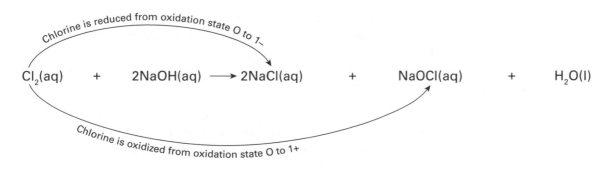

Chlorine is reduced from oxidation state O to 1−

$$Cl_2(aq) + 2NaOH(aq) \longrightarrow 2NaCl(aq) + NaOCl(aq) + H_2O(l)$$

Chlorine is oxidized from oxidation state O to 1+

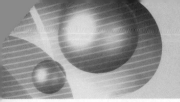

2.01 An introduction to organic chemistry

Key terms

Displayed formula: The relative positioning of atoms and the bonds between them. The displayed formula of propan-2-ol:

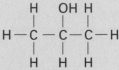

propan-2-ol

Empirical formula: The simplest whole-number ratio of atoms of each element present in a compound.

Functional group: The group of atoms responsible for the characteristic reactions of a compound. The functional group in a propan-2-ol molecule is the hydroxyl group, OH.

General formula: The simplest algebraic formula of a member of a homologous series. The general formula for alkenes is C_nH_{2n}.

Homologous group: A series of organic compounds having the same functional group but with each successive member differing by CH_2.

Molecular formula: The actual number of atoms of each element in a molecule.

Skeletal formula: The simplified organic formula, shown by removing hydrogen from alkyl chains, leaving just the carbon skeleton and associated functional groups. The skeletal formula of propan-2-ol:

propan-2-ol

The skeletal formula for cyclohexane, C_6H_{12}:

cyclohexane

The skeletal formula for benzene, C_6H_6:

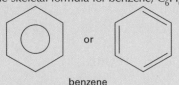

or

benzene

Structural formula: The minimal detail that shows the arrangement of the atoms in a molecule. The structural formula for propane is $CH_3CH_2CH_3$. The carboxyl group is shown as COOH. The ester group is represented by COOR.

The chemistry of carbon

Organic chemistry is the study of compounds containing carbon combined with other elements. Carbon is an unusual element, which has the ability to form chain and ring structures by joining carbon atoms together. This ability means that a very large number of organic compounds exist.

Organic compounds are grouped into 'families' which have the same molecular formula. Each of these families is called a **homologous series**.

You need to be familiar with three different homologous series:

- alkanes
- alkenes
- halogenoalkanes

Naming organic molecules

Functional groups

functional group	name
alkane	-ane
alkene	-ene
halogenoalkane	bromo-
	chloro-
	iodo-

Prefixes

number of C atoms	prefix
1	meth
2	eth
3	prop
4	but
5	pent
6	hex
7	hept
8	oct
9	non
10	dec

Side chains

number of C atoms	structure	Name
1	$-CH_3$	methyl
2	$-CH_2CH_3$	ethyl
3	$-CH_2CH_2CH_3$	propyl

The alkanes

The **alkanes** are **hydrocarbons**. This means they only contain carbon and hydrogen atoms.

They have single covalent bonds between the carbon atoms.

Alkanes have the general formula C_nH_{2n+2}

For example:

- The alkane with only 1 carbon atom has the formula CH_4.
- The alkane with 2 carbon atoms has the formula C_2H_6.
- The alkane with 26 carbon atoms has the formula $C_{26}H_{54}$.

The alkanes all have names ending in -ane.

For straight chain alkanes the beginning of the name tells you the number of carbon atoms in the chain.

- For example methane has one carbon atom in the chain; ethane has two carbon atoms, and so on.

If the chain is branched then it is said to have a side chain. The side chain is named methyl, ethyl, propyl, etc. depending on the number of carbon atoms it contains. This is then put at the front of the name of the main chain.

- For example methylbutane has a main chain of four carbon atoms and a side chain of one carbon atom.

If there is more than one position that the side chain could attach to the main chain then it is given a number. This is counted from the end of the chain in order to make the number of the side chain as small as possible.

- For the methylbutane molecule below right, a 2 is placed in front of the methyl group making 2-methylbutane.

If more than one side group of the same kind is attached to the chain then the prefix di-, tri-, etc. is used in front of the name of the side chain.

For each of the alkane molecules three types of formulae have been given: **displayed**, **structural**, and **molecular**. Make sure you are confident about what these show, as well as **skeletal formulae**, and that you can write and interpret each one.

The alkenes

The **alkenes** are also hydrocarbons, like the alkanes.

They have a double covalent bond between two of the carbon atoms. Alkenes have the general formula C_nH_{2n}

For example:

- The alkene with 2 carbon atoms has the formula C_2H_4.
- The alkene with 3 carbon atoms has the formula C_3H_6.
- The alkene with 12 carbon atoms has the formula $C_{12}H_{24}$.

The alkenes are named in the same way as the alkanes except that their name ends in -ene. The position of the C=C double bond is numbered if there is more than one place where it can go in a molecule. The numbering is such that it has the smallest number possible.

The halogenoalkanes

The **halogenoalkanes** are compounds containing carbon, hydrogen, and a halogen (fluorine, chlorine, bromine, or iodine) atom. Halogenoalkanes have the general formula $C_nH_{2n+1}X$ where X is the halogen atom.

For example:

- Chloromethane is CH_3Cl since it has one chlorine atom and one carbon atom.
- Bromoethane is C_2H_5Br as it has one bromine atom and two carbon atoms.

If the halogenoalkane has more than one of the same halogen atom then it is numbered di-, tri-, tetra-, etc. as with the side groups in alkanes.

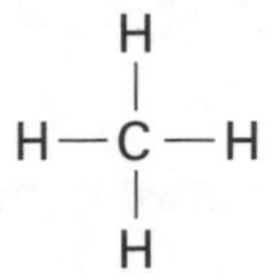

methane
CH_4

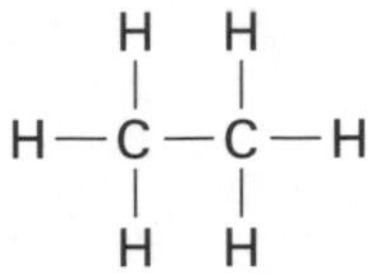

ethane
CH_3CH_3
C_2H_6

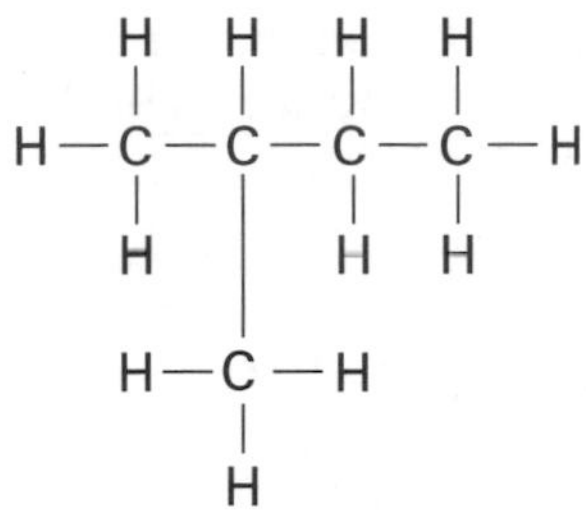

2-methylbutane
$CH_3CH(CH_3)CH_2CH_3$
C_5H_{12}

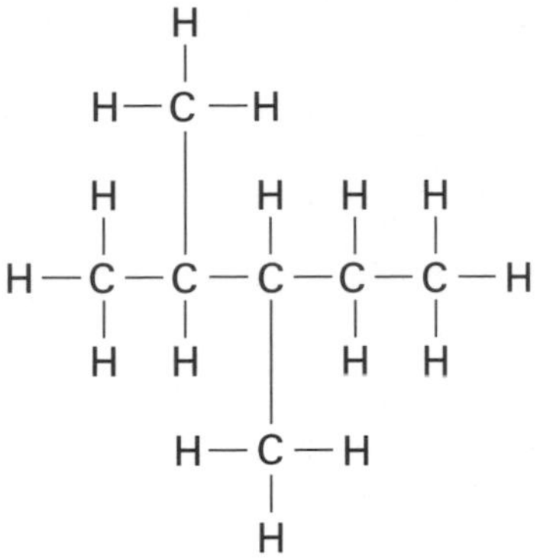

2,3-dimethylpentane
$CH_3CH(CH_3)CH(CH_3)CH_2CH_3$
C_7H_{16}

Alkanes

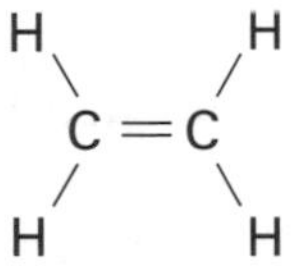

ethene
$CH_2 = CH_2$
C_2H_4

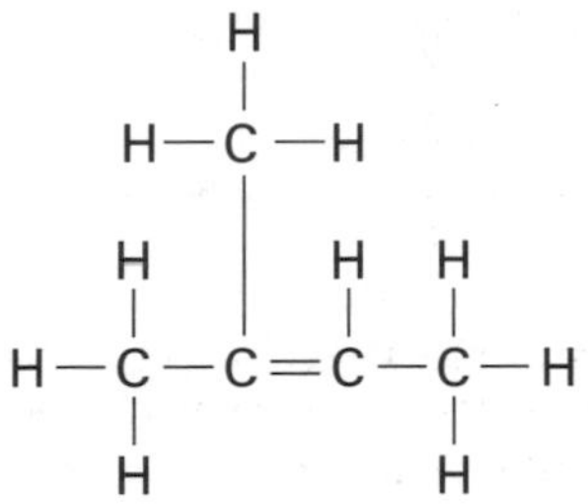

2-methylbut-2-ene
$CH_3C(CH_3)CHCH_3$
C_5H_{10}

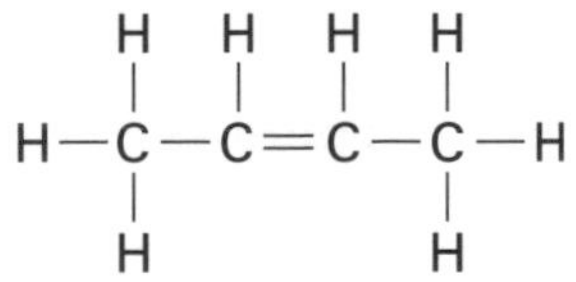

but-2-ene
$CH_3CH=CHCH_3$
C_4H_8

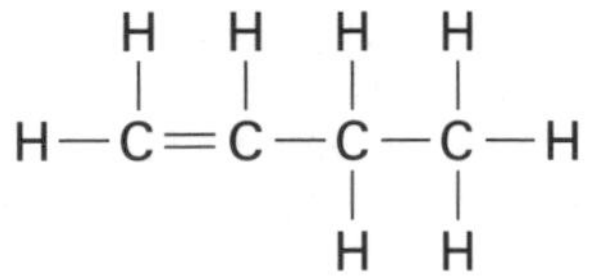

but-1-ene
$CH_2=CHCH_2CH_3$
C_4H_8

Alkenes

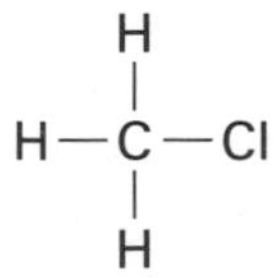

chloromethane
CH_3Cl

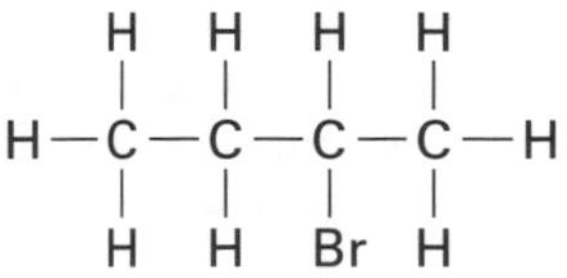

2-bromobutane
$CH_3CH_2CHBrCH_3$
C_4H_9Br

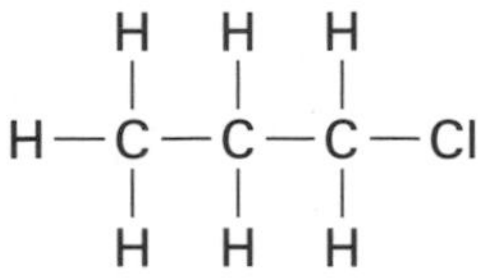

1-chloropropane
$CH_3CH_2CH_2Cl$
C_3H_7Cl

Halogenoalkanes

2.02 Isomerism

Isomerism occurs when two or more organic molecules have the same molecular formula but different arrangements of atoms. There are two main types of isomerism: **structural isomerism** and **stereoisomerism**.

Structural isomerism

Structural isomerism occurs when two or more molecules have the same molecular formula but different structural formulae.

For example, two or more molecules may have the same molecular formula but different arrangements of the carbon chain. Typically this will involve straight chain and branched chain structures.

Taking the example of C_4H_{10}, carbon atoms can be arranged in a straight chain forming butane or as a chain of three carbon atoms with one side chain forming methylpropane.

butane

methylpropane

For C_5H_{12} there are three possible chain isomers; the carbons atoms can be arranged in a straight chain forming pentane, as a chain of four atoms with a side chain forming methylbutane or as a chain of three atoms with two side chains forming dimethylpropane (the prefix di is used to show that there are two methyl groups attached to the central carbon atom).

Skeletal and displayed formulae showing isomers of C_5H_{12}

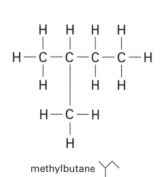

methylbutane

pentane

dimethylpropane

Another type of isomerism occurs when a functional group can be in more than one position on the carbon chain.

You must consider this when the molecule contains a chain of four or more carbon atoms or when the functional group is a halogen which can be in different positions on the chain.

- In butene (C_4H_8) there are two possible positions for the C=C, between atoms one and two or between atoms two and three.
- The positions are numbered so that the numbers are as small as possible.

In bromopropane (C_3H_7Br) the bromine atom can be bonded to the end carbon atom or the middle carbon atom, giving two possible isomers.

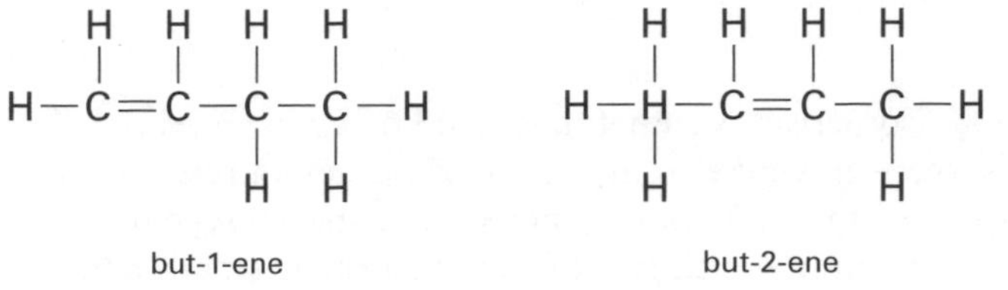

but-1-ene

but-2-ene

Structural isomers of C_4H_8

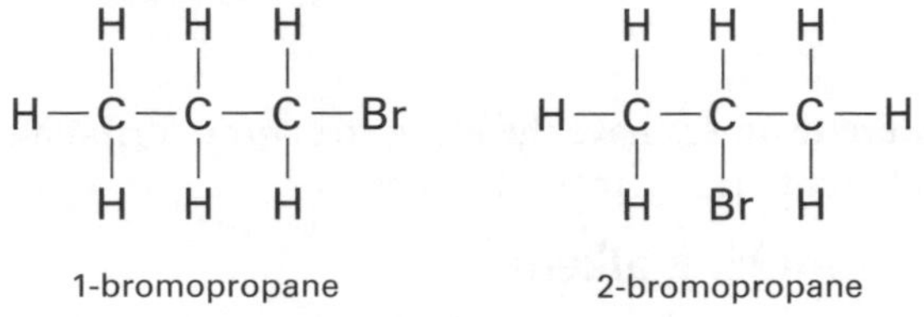

1-bromopropane

2-bromopropane

Structural isomers of C_3H_7Br

More complex examples

In an examination you are unlikely to be asked to draw a lot of isomers as it is a very time-consuming process! However, it is important that you approach drawing isomers in a systematic way to ensure that you don't miss any out. Name each one when you have drawn it to make sure that you do not draw the same molecule twice. For C_5H_{10} there are five possible isomers.

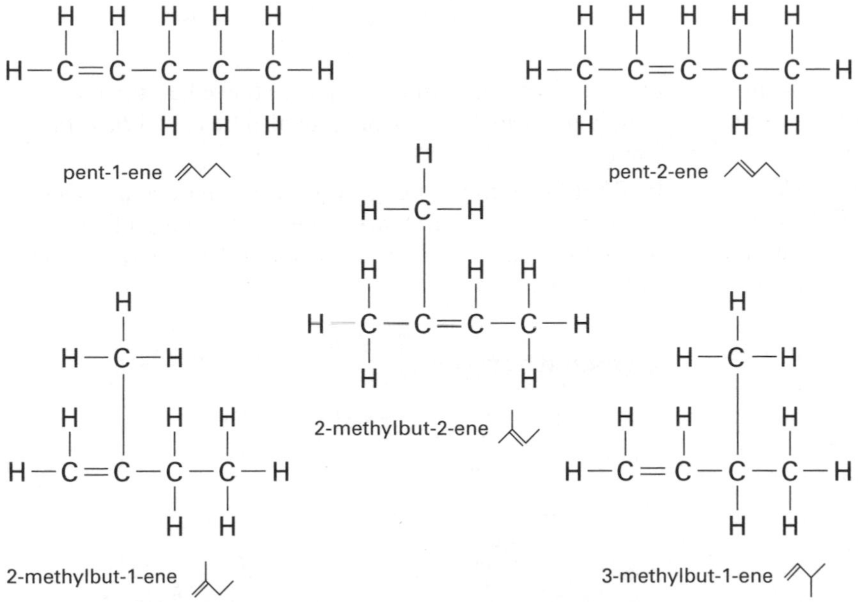

pent-1-ene

pent-2-ene

2-methylbut-2-ene

2-methylbut-1-ene

3-methylbut-1-ene

Naming more complex molecules

Organic compounds can become very complex but they are always named using the same systematic rules. These are sometimes called the IUPAC rules (IUPAC is the International Union of Pure and Applied Chemistry) as IUPAC standardizes the spelling of elements and the rules for naming organic compounds.

An example of a more complex compounds is shown below. You will meet many more during the course of your AS level studies.

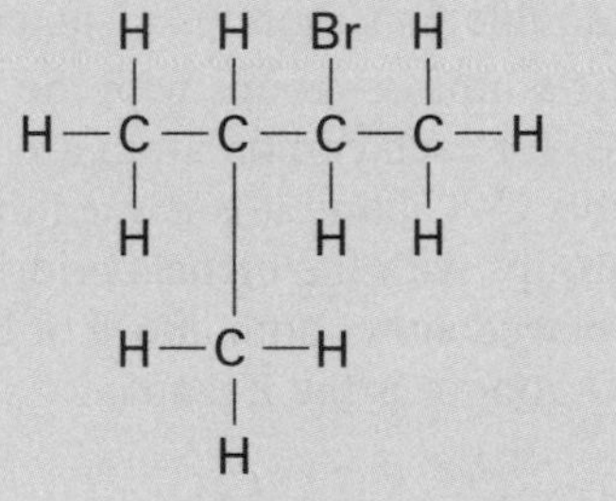

2-bromo-3-methylbutane 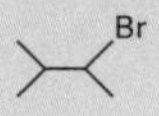
$CH_3CH(CH_3)CHBrCH_3$
$C_5H_{11}Br$

Determining molecular formula

The molecular formula of a compound is a whole number multiple of its empirical formula. The empirical formula can be calculated from percentage composition data and the molecular mass determined from a mass spectroscopy experiment.

An organic compound is analysed and found to have the following percentage composition by mass: carbon – 82.76%, hydrogen – 17.24%. The mass spectrum shows a molecular ion peak at a mass/ charge value of 58.

element	C	H
% by mass	82.76	17.24
÷ A_r	6.897	17.24
÷ smallest	1	2.5
× 2	2	5

Empirical formula C_2H_5

The empirical formula is calculated in the same way as for an inorganic compound. Note that the final step involves multiplying the data by two in order to obtain a whole number ratio.

The mass of the empirical formula is calculated $(2 \times 12) + (5 \times 1) = 29$

The molecular mass is then divided by the empirical mass $58/29 = 2$

This tells you that the molecular formula is twice the empirical formula.

The molecular formula is C_4H_{10}.

Questions

1 Draw displayed formulae and name all the structural isomers of C_6H_{14}.

2 Draw skeletal formulae and name the structural isomers of C_4H_9I.

3 Draw displayed formulae and name all the isomers of C_6H_{12}. Remember to consider both the carbon chain and the functional group.

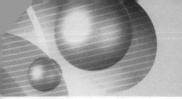

2.03 Stereoisomers

Stereoisomers have the same structural formula but a different arrangement in space.

Introducing alkenes

Alkenes are **unsaturated** hydrocarbons with the general formula C_nH_{2n}.

Alkenes have a C=C double bond which consists of a **sigma bond** and a **pi bond**. The pi bond forms above and below the axis of the carbon atoms.

Ethene is a **planar** (flat) molecule because of the C=C arrangement in alkene molecules.

The pi bond has a high electron density. Electrophiles can attack this high electron density.

As a result alkenes are more reactive than alkanes.

E/Z stereoisomers

Rotation of the double C=C bond would require the pi bond to be broken. This requires a lot of energy so there is restricted rotat ion about the double C=C bond.

If the two carbon atoms involved in the double C=C bond have different groups attached to them, *E/Z* isomers are formed. In the *E/Z* naming system each of the atoms or groups of atoms attached to the carbons involved in the double C=C bond is assigned a priority.

If the atoms or groups with the highest priority (greatest mass) for each carbon atom are arranged across the double C=C bond it is named the *E* isomer. If the atoms or groups with the highest priority for each carbon atom are placed either both above or both below the double C=C bond it is the *Z* isomer.

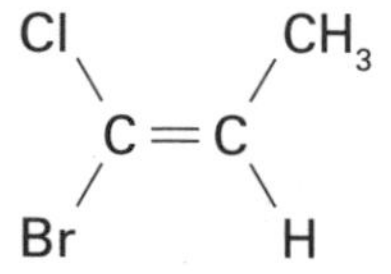

E-1-bromo-1-chloropropene

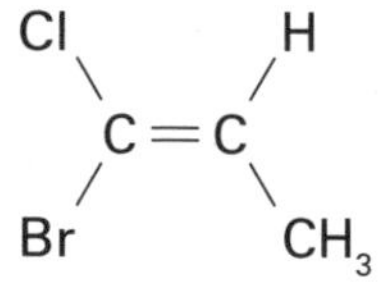

Z-1-bromo-1-chloropropene

As a result 1-bromo-1-chloropropene exists as two stereoisomers. These isomers are not mirror images of each other.

In 1-bromo-1-chloropropene the bromo group is given a higher priority than the chloro group and the methyl group is given a higher priority than the hydrogen group.

Notice that in the *E* form the bromo and the methyl groups are arranged on opposite sides of the double C=C bond. The *E* comes form the German word ***entegen*** which means *opposite*.

In the *Z* form the bromo group and the methyl group are arranged on the same side of the double C=C bond. The *Z* comes from the German word ***zusammen*** which means *together*.

E/Z isomerism is sometimes also known as *cis-trans* isomerism, where *cis* means the *Z* isomer and *trans* means the *E* isomer. *Cis-trans* isomerism is a special type of *E/Z* isomerism in which two of the substituent groups bonded to the double C=C bond are the same.

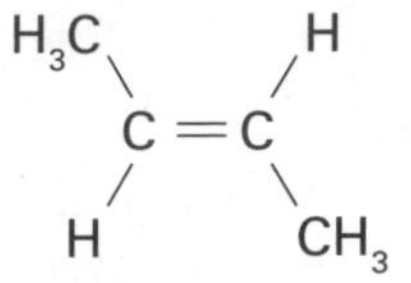

E-but-2-ene
(*trans*)

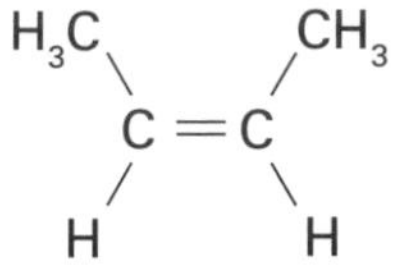

Z-but-2-ene
(*cis*)

In *E*-but-2-ene the same groups are placed across the double bond so *E*-but-2-ene may also be called *trans*-but-2-ene.

In *Z*-but-2-ene the same groups, rather than being placed across the double bond, are either both on top of or both below the double C=C bond so *Z*-but-2-ene may also be called *cis*-but-2-ene.

Worked example

Alkenes A, B, C, and D are shown below.

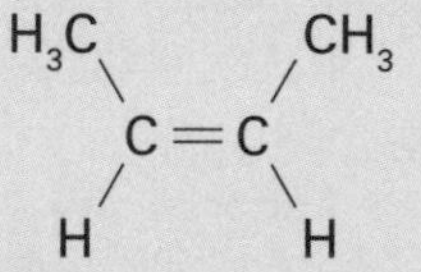

A

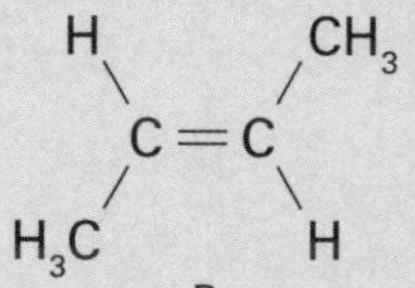

B

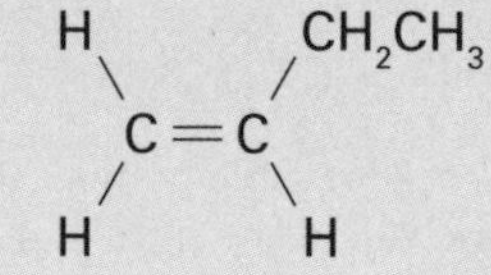

C

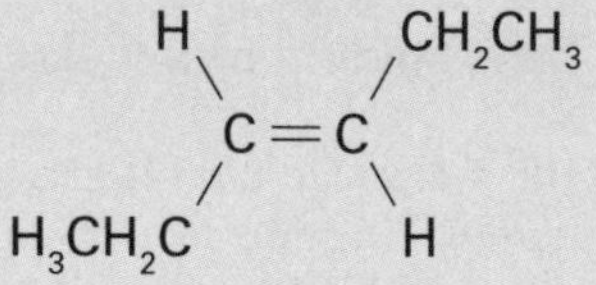

D

a Explain why alkene C does not show *E/Z* isomerism.

b Which two alkenes (A, B, C, or D) are a pair of *E/Z* isomers?

Answers

a In alkene C the two groups attached to one of the carbon atoms involved in the double C=C bond are the same.

b A and B are *E/Z* isomers.

2.04 Atom economy

What is atom economy?

The **atom economy** of a chemical reaction is the proportion of reactants that are converted into useful products.

- Processes with high atom economy are more efficient and produce less waste.
- This is important for sustainable development.

Atom economy is calculated by looking at the mass of desired product in relation to the total mass of the reactants or products. These may be expressed as mass in grams or as relative molecular mass.

Calculating atom economy

The reaction of methane with steam produces hydrogen gas along with carbon monoxide as a waste product. Calculate the percentage atom economy in relation to hydrogen gas.

$$CH_4(g) + H_2O(g) \rightarrow 3H_2(g) + CO(g)$$

% atom economy

$$= \frac{\text{molecular mass of desired product}}{\text{sum of molecular masses of all products}} \times 100$$

$$= \frac{3 \times 2}{((3 \times 2) + 28)} \times 100 = \frac{6}{34} \times 100 = 17.6\%$$

Note that the ratios of each reacting species have been taken into account here.

Optimizing atom economy

A suitable catalyst may also allow chemists to select a different reaction to use. If the chosen reaction has a higher atom economy then the process will be more efficient and less waste will be produced.

In addition reactions the reactant molecules add together to form the desired product. This means that addition reactions have an atom economy of 100%.

In substitution reactions the atom economy is less than 100% as other products are also made. Some reactions may have:

- a low atom economy (the reaction chosen produces the desired product plus lots of other waste products)
- but also a high percentage yield (the actual yield of the desired product relative to the theoretical yield of the desired product is high).

For better sustainability in the future chemists should choose reactions which have a high atom economy.

Percentage yield of a reaction

The **percentage yield** is the actual amount in moles of the product shown as a percentage of the expected yield in moles.

$$\% \text{ yield} = \frac{\text{actual amount of product in moles}}{\text{theoretical amount of product in moles}} \times 100\%$$

Worked example

In this reaction ethanol reacted with ethanoic acid to make the ester ethyl ethanoate and water.

$$C_2H_5OH + CH_3COOH \rightarrow CH_3COOC_2H_5 + H_2O$$

2.3 g of ethanol were used with an excess of ethanoic acid. The reaction produced 2.2 g of ethyl ethanoate. What is the percentage yield of the reactions?

$$\text{moles of ethanol used} = \frac{\text{mass}}{\text{molar mass}}$$

$$= \frac{2.3 \text{ g}}{46 \text{ g mol}^{-1}}$$

$$= 0.050 \text{ mol}$$

This reaction should produce 0.05 mol of ethyl ethanoate. The theoretical yield is 0.05 mol.

Actual amount of ethyl ethanoate made in the reaction

$$= \frac{\text{mass}}{\text{molar mass}}$$

$$= \frac{2.2 \text{ g}}{88 \text{ g mol}^{-1}}$$

$$= 0.025 \text{ mol}$$

Percentage yield $= \frac{0.025}{0.050} \times 100\%$

$$= 50\%$$

Questions

1 Chlorine gas can be obtained from the electrolysis of brine. The equation for this process is:

$$2NaCl(aq) + 2H_2O(l) \rightarrow$$
$$2NaOH(aq) + Cl_2(g) + H_2(g)$$

Calculate the atom economy for producing chlorine.

2 The Haber Process for making ammonia by reacting nitrogen and hydrogen gases

$$N_2(g) + 3H_2(g) \rightleftharpoons 2NH_3(g)$$

typically has a percentage yield of around 15%. Explain what is meant by the term percentage yield and compare this to the atom economy for this process.

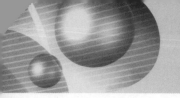

2.05 The alkanes

Hydrocarbons

Hydrocarbons are molecules containing carbon and hydrogen only.

The alkanes can be defined as:

- saturated hydrocarbons (they contain C–C single bonds)
- molecules with the general formula C_nH_{2n+2}
- Alkanes and cycloalkanes are saturated hydrocarbons. They do not contain double C=C bonds.
- In alkane molecules each carbon atom is surrounded by four pairs of electrons. These pairs of electrons repel each other leading to a tetrahedral shape around each carbon atom.

tetrahedral

methane

- Methane has one carbon atom surrounded by four pairs of electrons so a methane molecule has a tetrahedral shape and a bond angle of 109.5°.

Straight chain and branched chain alkanes

For alkanes with more than three carbon atoms it is possible to draw the molecules as straight or as branched chains:

- These are called chain isomers.
- Their chemistry will be very similar.
- They have slight differences in physical properties such as volatility (the ease with which a substance evaporates), melting point, and boiling point.
- These properties also vary with increasing chain length.

Boiling point of the alkanes

Look at the graph showing number of carbon atoms (i.e. chain length) and boiling point.

- As the carbon chain gets longer the boiling point increases.
- The difference between each molecule is one $-CH_2-$ unit.
- As a result the trend is almost linear.

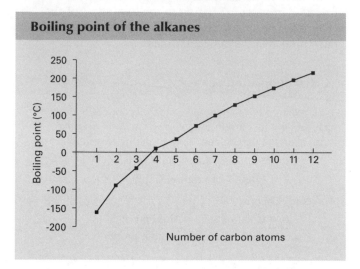

Boiling point of the alkanes

Explanation for this trend:

- Remember that melting point and boiling point trends are related to the forces between the molecules.
- For the alkanes (which are non-polar molecules) the intermolecular force is a van der Waals' force.
- The strength of this force increases with increasing chain length as the molecules are bigger.
  - As a result the boiling points increase.

Now look at the data (below) showing the boiling points of the isomers of C_5H_{12}:

- Note that the boiling point decreases with increased chain branching.
- This is because the surface area of contact of the molecule is decreasing.
  - As a result the van der Waals' force is weaker.

pentane
C_5H_{12}
boiling point = 36°C

methylbutane
C_5H_{12}
boiling point = 28°C

dimethylpropane
C_5H_{12}
boiling point = 141°C

Fractional distillation and the petrochemical Industry

Petroleum (crude oil) is a mixture of hydrocarbons, most of them alkanes.

- These alkanes are separated into groups with similar boiling points.
- These groups are called **fractions**.
- The process of separating the petroleum is called **fractional distillation**.
- This relies on the differences in boiling points.

The fractionating column

You will not be asked to draw a fractionating column but it is important that you understand how it works.

- The petroleum is heated to around 350°C so that it forms a vapour.
- The vapour is then passed into the fractionating column.
- The temperature of the column falls as it is ascended (this is called a negative temperature gradient).
- The largest hydrocarbons have the highest boiling points so remain as liquids at the bottom of the column.
- These are tapped off as residue.
- The smallest hydrocarbons have the lowest boiling points so rise up through the column and leave the top as gases.
- The remaining hydrocarbons rise up the column until they reach the region that corresponds to their boiling point.
- They then condense at the bubble caps and are removed from the column as liquids.

Questions

1 Think about the alkanes with molecular formula C_6H_{14}.

   a Write full displayed formulae for all the possible isomers.

   b Name each one.

   c Arrange the isomers in order of increasing boiling point starting with the lowest.

   d Make sure you can explain the order you have chosen (look back at the page on intermolecular forces if you need to).

2 Write a three sentence explanation of how different hydrocarbon products are obtained from petroleum, and how they are used.

Uses of the fractions of petroleum

number of carbon atoms in fraction	typical range of boiling points (°C)	typical use of fraction
1–4	below 20	camping gas
4–12	20–125	fuel for cars
7–14	125–175	chemical feedstock
11–15	175–250	jet fuel and chemical feedstock
15–19	250–350	fuel for vehicles and heating
20–30	over 350	lubricating oil
30–40	350–500	fuel for power stations and ships

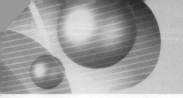
The catalytic converter

You will already know that a catalyst is a substance that changes the rate of a reaction but is chemically unaltered at the end of the reaction.

A catalytic converter is fitted to the exhaust system of a car.

- It removes pollutants from the exhaust gases before they are released through the end of the exhaust.
- The catalytic converter decreases the amount of carbon monoxide and nitrogen monoxide being released into the atmosphere. First these gases are adsorbed (held) on the surface of the catalyst. Then the reaction takes place. Finally the products are desorbed from the surface of the catalyst.
- Alloys of platinum, palladium, and rhodium are used in the catalytic converters fitted to modern cars. These alloys catalyse the reactions between CO and NO thereby reducing the levels of these gases being released into the atmosphere.

$$2CO + 2NO \rightarrow 2CO_2 + N_2$$

Complete and incomplete combustion

Alkanes are widely used as fuels in industry, in the home, and in transport.

Complete combustion

For **complete combustion** to occur:
- An unlimited supply of oxygen is needed.
- Alkane + oxygen → carbon dioxide + water + release of energy

For methane (the main fossil fuel in natural gas):

$$CH_4(g) + 2O_2(g) \rightarrow CO_2(g) + 2H_2O(l)$$

For octane, a longer chain hydrocarbon, the products are identical.

$$C_8H_{18}(g) + 12\tfrac{1}{2}O_2(g) \rightarrow 8CO_2(g) + 9H_2O(l)$$

Take care when balancing equations like this. Often a *half* is used to balance the equation.

Incomplete combustion

In a **limited supply** of oxygen, incomplete combustion occurs.
- The hydrogen in the hydrocarbon still forms water.
- The carbon only undergoes partial oxidation to form carbon monoxide.
- In some cases unburnt carbon particles are released as soot during incomplete combustion.
- Carbon monoxide is a toxic gas that binds to the haemoglobin in red blood cells and prevents them carrying oxygen.
- Carbon monoxide is difficult to detect because it is colourless, odourless, and tasteless.

The incomplete combustion of the same fuels as shown above gives:

$$CH_4(g) + 1\tfrac{1}{2}O_2(g) \rightarrow CO(g) + 2H_2O(l)$$

$$C_8H_{18}(g) + 8\tfrac{1}{2}O_2(g) \rightarrow 8CO(g) + 9H_2O(l)$$

- Incomplete combustion can occur in car engines or in badly ventilated gas fires or boilers in the home.

Inside an engine

The combustion temperature of fuel in an engine can be 1000 °C.
- Under these conditions enough energy is available for the nitrogen and the oxygen in air to react together forming a number of oxides of nitrogen.
- These are referred to in general as NO_x.

One of the gases formed is nitrogen monoxide which can undergo further **oxidation** in the air to nitrogen dioxide.

$$N_2(g) + O_2(g) \rightarrow 2NO(g)$$

$$2NO(g) + O_2(g) \rightarrow 2NO_2(g)$$

Some of the fuel passes through the car engine without undergoing oxidation. These gases are called unburned hydrocarbons or volatile organic compounds.

In sunlight these unburned hydrocarbon compounds can react with nitrogen oxides to form low-level ozone which is a major component of photochemical smog.

Global warming

Our atmosphere is made up of several gases, a number of which play an important role in controlling our climate.
- Some of these gases are described as **greenhouse gases** eg water vapour, carbon dioxide, and methane.

The C=O, O-H and C-H bonds are good at absorbing infrared radiation and re-emitting it in a different direction, which warms up the atmosphere. This effect is called the greenhouse effect. The contribution that a particular gas makes toward the greenhouse effect depends on:

- the concentration of the gas in the atmosphere
- the gas's ability to absorb infrared radiation.

The greenhouse effect keeps the planet warm enough to sustain life. However, processes such as the burning of fossil fuels have released large amounts of greenhouse gases.

As a result our planet is getting warmer. This is **global warming** and can lead to climate change.

Chemists provided society with evidence that established that global warming was taking place. They noticed that areas of permafrost and ice sheets were melting and that storms and flooding were becoming more severe and happening more frequently. It is important that global warming is controlled.

In the future, if global warming continues it could lead to a change in weather patterns leading to more droughts and water shortages in some areas while heavier rainfall could cause flooding in other places.

In addition scientists fear that storms and hurricanes could become more violent and occur more frequently. Sea levels may also rise. These changes could lead to widespread and serious problems. Chemists are actively involved in minimizing the effect of climate change.

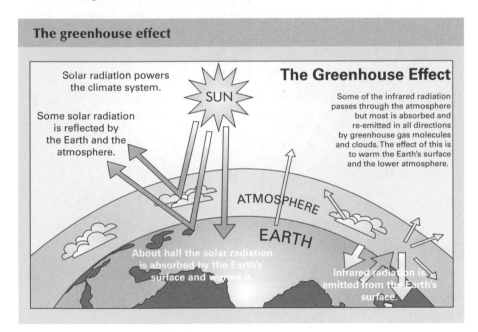

The greenhouse effect

Carbon capture and storage (CCS)

Chemists are developing ways of reducing the amount of carbon dioxide being released into the atmosphere by capturing some of the carbon dioxide released by power stations and then storing it safely.

- Some carbon dioxide could be stored in geological formations such as the porous rocks of old oil and gas fields.
- Some carbon dioxide could be stored as a liquid deep beneath the oceans.
- Some carbon dioxide could also be reacted with metal oxides to form carbonate rocks:

$$CO_2(g) + CaO(s) \rightarrow CaCO_3(s)$$

The resulting metal carbonate would be very stable and easy to store.

The Kyoto protocol

In 1997 over 100 countries agreed to sign up to the Kyoto protocol. This initiative committed countries to reduce their emissions of greenhouse gases. Chemists monitor each country's progress towards its targets.

Questions

1 Camping gas is a mixture of the alkanes propane and butane.

  a Write balanced equations for the complete combustion of butane and propane.

  b Write balanced equations for the incomplete combustion of butane and propane.

2.07 Catalytic cracking

Hexane, cyclohexane, and benzene

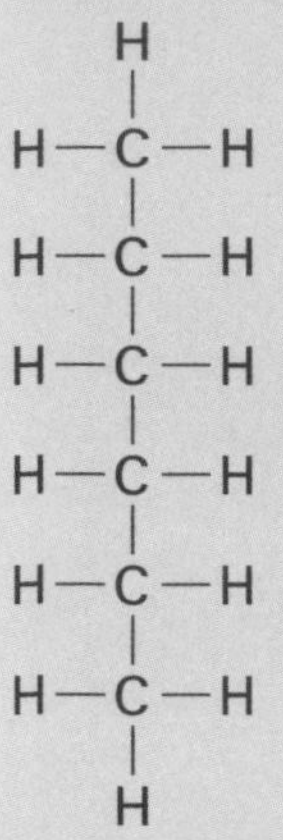

hexane

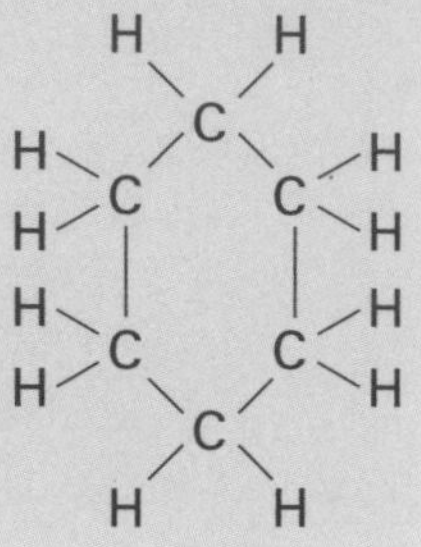

cyclohexane

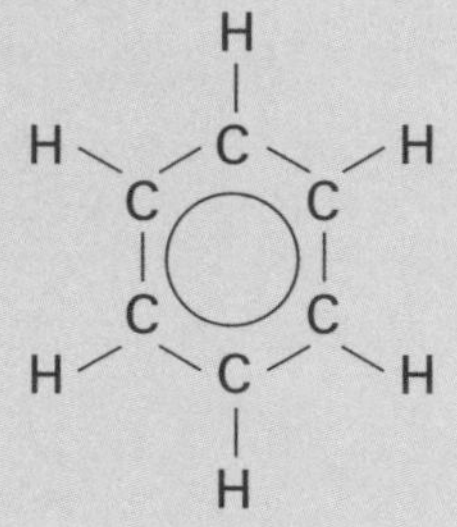

benzene

Why cracking?

The **fractional distillation** of petroleum produces a range of fractions with different boiling points. The mixture of fractions obtained often contains a higher proportion of heavier fractions than is needed. The **petrochemical industry** carries out the processing of these heavier fractions in order to make more useful ones.

- **Cracking** is a decomposition reaction which involves the breaking of C–C single bonds.
- This results in the formation of shorter chain alkanes and alkenes.

The heptane molecule can be cracked in a number of ways; in each equation note that both an alkane and an alkene are formed.

$$C_7H_{16} \rightarrow C_5H_{12} + C_2H_4$$
$$C_7H_{16} \rightarrow C_4H_{10} + C_3H_6$$
$$C_7H_{16} \rightarrow C_3H_8 + 2C_2H_4$$
$$C_7H_{16} \rightarrow C_5H_{10} + C_2H_4 + H_2$$

The shorter chain alkanes are more useful as fuels than the original alkane as they are more **volatile** (evaporate more readily) and burn more easily. The alkenes are used as raw materials for making polymers.

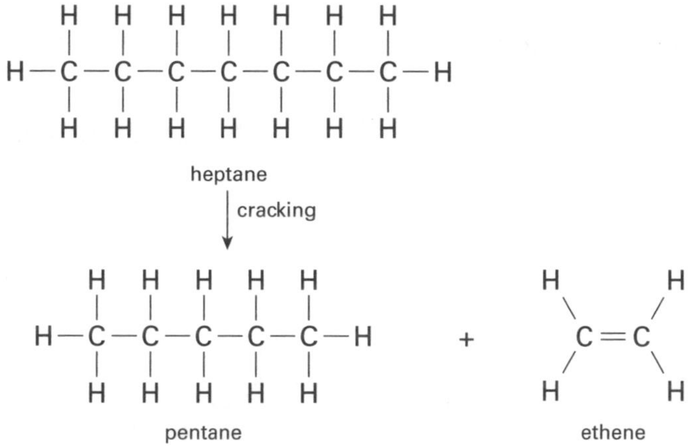

heptane

cracking

pentane   +   ethene

Catalytic cracking details

	catalytic cracking
raw material	long chain alkane
temperature	500°C
pressure	slightly above atmospheric pressure
catalyst	silica and aluminium oxide or zeolite
products	aromatic hydrocarbons
uses of products	motor fuels (short chains burn more readily, more volatile)
notes	• more efficient than thermal cracking • produces more branched, cyclic, and aromatic hydrocarbons

Petrol

Petrol should ignite in an engine because of a spark at just the right time. If it ignites too soon a knocking sound is recorded, which reduces efficiency of combustion.

The higher the octane rating of the petrol the less it is likely to cause knocking.

Economic reasons for cracking

The fractional distillation process produces disproportionate amounts of some of the fractions of petroleum.

- The amount of petrol produced is not enough for our needs.
- Too much of the naphtha fraction is produced.

The cracking process helps us make maximum use of petroleum by converting the less useful fractions into more useful ones. This is an essential process, as petroleum is a **non-renewable** resource.

Petrol

Petrol is a mixture of hydrocarbons:

- Petrol manufacturers mix hydrocarbons to obtain the petrol with the right properties.
- The petrol must vaporize readily enough to ignite in the engine.
- It must not be so volatile that excess vapour enters the engine.
- It must not ignite in the cylinders of the engine too early.

These properties are temperature dependent.

- For example, longer chains are not so volatile so are used when the external temperature is warmer.
- Winter blends would vaporize too readily in the summer.

You may have heard of the term **octane rating**. This tells us how easily the petrol ignites. A lower octane rating indicates that a fuel will more readily auto-ignite. This means that the petrol ignites in the engine before it should. This causes a knocking sound and can lead to engine damage.

Processing of hexane forms three different molecules, each of which has a unique octane rating (see table opposite):

You do not need to remember these values but you do need to have an understanding of the difference between a low and a high octane rating.

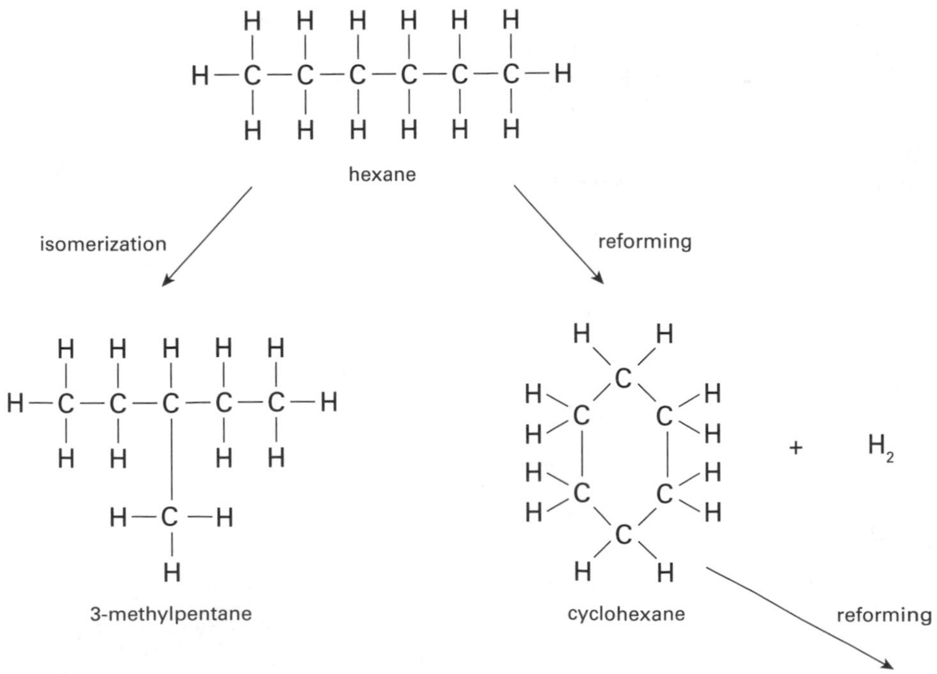

name of molecule	octane rating
hexane	19
3-methylpentane	86
cyclohexane	83
benzene	106

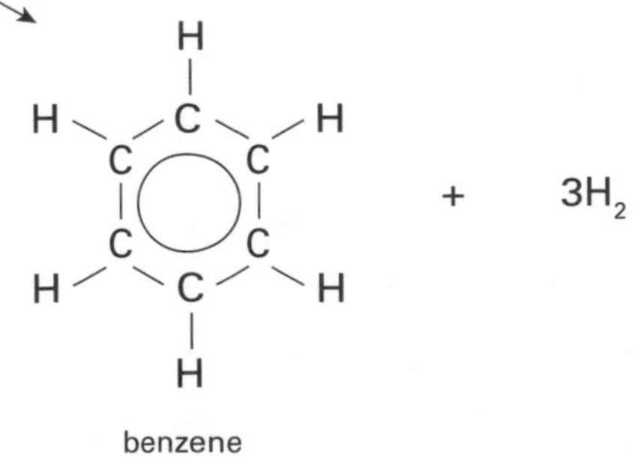

> **Questions**
>
> 1 Write equations using displayed formulae for the cracking of:
>   a hexane, C_6H_{14}
>   b octane, C_8H_{18}
> 2 Explain in your own words the following terms:
>   a octane rating
>   b auto-ignition
>   c volatility

hexane

isomerization   reforming

3-methylpentane   cyclohexane   reforming

In the petroleum industry straight-chain hydrocarbons (which do not burn efficiently) are processed into branched alkanes and cyclic hydrocarbons. These fuels undergo more efficient combustion and are therefore better fuels.

benzene

57

Radicals

- **Radicals** are very reactive species. They are species which have an unpaired electron.
- They are formed when a covalent bond breaks so that one electron is transferred to each of the atoms.
- This is called **homolytic fission** and forms two radicals.

$$A-B \rightarrow A\cdot + B\cdot$$

Notice that the dot, which represents the unpaired electron, is written next to the atom which has the unpaired electron.

Worked example

a Explain what a radical is.

b Write an equation to show how a chlorine molecule can form chlorine radicals.

Answer

a A radical is a species which has an unpaired electron.

b $Cl_2 \rightarrow 2Cl\cdot$

The radical substitution of methane

Methane reacts with chlorine by a radical substitution reaction.

We can sum up this reaction using the equation

$$CH_4 + Cl_2 \rightarrow CH_3Cl + HCl$$

Initiation

- UV radiation provides the energy required for the homolytic fission of the chlorine molecule.

$$Cl_2 \rightarrow 2Cl\cdot$$

Propagation

- Next the chlorine radical reacts with methane to form hydrogen chloride and a methyl radical.

$$Cl\cdot + CH_4 \rightarrow HCl + \cdot CH_3$$

- Then the methyl radical reacts to form chloromethane and a chlorine radical.

$$\cdot CH_3 + Cl_2 \rightarrow CH_3Cl + Cl\cdot$$

Notice that as the radical reacts it forms a new molecule and another radical.

Other products include

dichloromethane

trichloromethane

tetrachloromethane

Termination

In the termination step two radicals react together to form a new molecule.

$$Cl\cdot + Cl\cdot \rightarrow Cl_2$$
$$\cdot CH_3 + Cl\cdot \rightarrow CH_3Cl$$
$$\cdot CH_3 + \cdot CH_3 \rightarrow C_2H_6$$

Notice that the three steps in the radical substitution of methane are initiation, propagation, and termination.

- In the **initiation** step radicals are made.
- In the **propagation** steps radicals react with molecules to form new molecules and radicals.
- In the **termination** steps two radicals join together. A variety of new molecules are formed.

Worked example

In the presence of UV radiation methane reacts with chlorine.

a Why does the reaction require the presence of UV radiation?

b State the formula of a product of the reaction which does not contain chlorine.

Explain how this product is formed.

Answer

a The UV radiation is required to split the chlorine molecule into two chlorine radicals.

b C_2H_6. It is formed by the termination step involving two methyl radicals.

$$\cdot CH_3 + \cdot CH_3 \rightarrow C_2H_6$$

The use of radical substitutions in synthesis is limited because further substitutions may result in a mixture of products. Alkanes can also be substituted by bromine.

Uses of chloroalkanes and chlorofluoroalkanes

- **Chloroalkanes** are alkane molecules which have one or more hydrogen atoms substituted by chlorine atoms.
- Chloroalkanes are good solvents.
- **Chlorofluoroalkanes**, CFCs, are alkanes that have hydrogen atoms substituted by chlorine and fluorine atoms.
- CFCs have been used as propellants in aerosols and as refrigerants.
- CFCs have been used as blowing agents in the manufacture of expanded polymers such as polystyrene foam. The use of CFCs has been phased out and carbon dioxide is now used instead.
- CFCs have also been used in air-conditioning.

CFCs were used because of their low reactivity, volatility and because they were non-toxic.

- CFCs can escape into the Earth's atmosphere where they can damage the ozone layer.

Benefits of the ozone layer

- **Ozone**, O_3, is formed in the upper atmosphere.
- UV light from the Sun provides the energy required to break the covalent bond in an oxygen molecule.

$$O_2 \rightarrow O + O$$

- The oxygen atoms then react with oxygen molecules to form ozone.

$$O + O_2 \rightarrow O_3$$

- Ozone helps to filter out harmful UV radiation.

This is important because over-exposure to UV radiation can cause skin cancers, cataracts, and damage to crops. Ozone molecules can be broken down by ultraviolet light to produce oxygen molecules and an oxygen radical.

$$O_3 \rightarrow O_2 + O\bullet$$

In this way the ozone molecules in the ozone layer are continually being formed and destroyed. An equilibrium is set up in which the concentration of ozone remains fairly constant over time.

$$O_2 + O\bullet \rightleftharpoons O_3$$

Decomposition of ozone

- In the upper atmosphere the CFC molecules decompose to form radicals.

$$CCl_3F \rightarrow \bullet CCl_2F + Cl\bullet$$

- The chlorine radical then removes ozone forming the chlorine oxide radical and oxygen.

$$Cl\bullet + O_3 \rightarrow ClO\bullet + O_2$$

The chlorine oxide radical can also react with oxygen atoms to form oxygen molecules.

$$ClO\bullet + O \rightarrow Cl\bullet + O_2$$

The chlorine radical Cl• is regenerated in the second step, so the chlorine radical acts as a catalyst for the decomposition of ozone.

Nitrogen oxides NOx are made when nitrogen and oxygen atoms react together at high temperatures and pressures inside car and aircraft engines and in thunderstorms. The nitrogen oxides break down to form the nitrogen monoxide radical NO•.

This nitrogen monoxide radical catalyses the decomposition of ozone:

$$\bullet NO + O_3 \rightarrow \bullet NO_2 + O_2$$
$$\bullet NO_2 + O \rightarrow \bullet NO + O_2$$

Notice how the nitrogen monoxide radical is regenerated in the second step of the reaction.

- Many old refrigerators and freezers contain CFCs. When they are disposed of the CFCs must first be removed to prevent the gases escaping into the atmosphere.

Legislation and new compounds

- Scientists have helped us to understand how the hole in the ozone layer has formed and the legislation to ban CFCs has been widely supported by scientists.
- Scientists have also developed alternative compounds including hydrochlorofluorocarbons, HCFCs, which contain less chlorine, and HFCs which do not contain chlorine and are not thought to affect the atmosphere.

Worked example

a In the past CFCs were widely used. State a use of CFCs.

b Explain why the use of CFCs has been greatly reduced.

c Today HFCs have widely replaced CFCs in many applications. Why are HFCs not thought to affect the Earth's atmosphere?

Answer

a They were widely used as propellants in aerosols and as refrigerants.

b CFCs decompose to form radicals.

$$CCl_3F \rightarrow \bullet CCl_2F + Cl\bullet$$

These radicals then react with ozone.

$$Cl\bullet + O_3 \rightarrow ClO\bullet + O_2$$
$$ClO\bullet + O_3 \rightarrow 2O_2 + Cl\bullet$$

c HFCs do not contain chlorine.

Ozone and CFCs

Scientists first discovered that the level of ozone in the upper atmosphere was decreasing. They discovered that the 'hole' in the ozone layer is caused by CFCs which had escaped into the atmosphere.

Questions

1 How are radicals formed?
2 Name the type of reaction by which methane reacts with chlorine.
3 How are CFCs used?

Alkenes

Alkenes are unsaturated hydrocarbons with the general formula C_nH_{2n}. Adjacent p orbitals on the carbon atoms involved in the double bond overlap to form a **pi bond**. The pi bond forms above and below the axis of the single sigma bond.

In alkene molecules each of the carbon atoms involved in the double bond is surrounded by one double bond and two single bonds. The bonds repel each other leading to a trigonal planar shape around each of these carbon atoms.

Heterolytic fission

In heterolytic fission a bond is broken and both electrons from the bond go to the same atom or group of atoms. The Br-Br bond is broken with both the electrons going to the same bromine.

$$Br\!-\!Br \longrightarrow Br^+ + Br^-$$

Heterolytic fission of a covalent bond forms a cation and an anion.

Reactions of alkenes

In addition reactions two species join together.

An **electrophile** is a species that can accept a pair of electrons to form a new covalent bond.

Electrophiles can attack the high electron density of the pi bond in alkenes.

As a result alkenes undergo electrophilic addition reactions.

Electrophilic addition

In electrophilic addition reactions an electrophile is attracted to an electron-rich centre or atom (such as the electrons in the double bond of an alkene molecule). The electrophile accepts a pair of electrons to form a new covalent bond.

In reaction mechanisms a curly arrow represents the movement of a pair of electrons. A curly arrow can be used to show that a covalent bond is being broken or that a new bond is being formed. When drawing out reaction mechanisms make sure that all relevant dipoles are included.

Worked example
Define the term electrophile.
Answer
An electrophile is a species which can accept a pair of electrons to form a new covalent bond.

Reaction with bromine

Alkenes react with bromine to form dibromoalkanes.
Example

$$C_2H_4 + Br_2 \rightarrow C_2H_4Br_2$$

This is an electrophilic addition reaction.

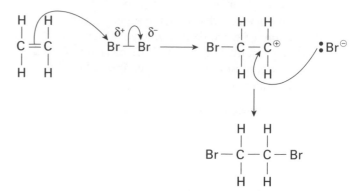

The reaction mechanism for the electrophilic addition of bromine to ethene.

Notice in the mechanism above how the electron pair in the pi bond of the ethene molecule is attracted to the slightly positively charged bromine.

Worked example
a Give the equation for the reaction between propene, C_3H_6, and bromine, Br_2, using structural formulae.
b What type of reaction is this?
Answer
a (structural formula equation)
b electrophilic addition

Reaction with hydrogen bromide

Alkenes also react with hydrogen bromide to form bromoalkanes.

Example

$$C_2H_4 + HBr \rightarrow C_2H_5Br$$

This is an electrophilic addition reaction.

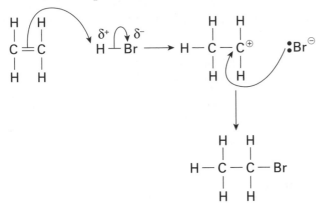

The reaction mechanism for the electrophilic addition of hydrogen bromide to ethene.

Reaction with hydrogen

Alkenes react with hydrogen gas in the presence of a nickel catalyst to form alkanes. A temperature of 150 °C is used.

$$C_2H_4 + H_2 \rightarrow C_2H_6$$

Testing for the presence of double C=C bonds

Bromine water can be used to test for the presence of double C=C bonds in molecules. First the orange coloured bromine water is added to the compound being tested. Then the mixture is shaken. If the molecule is unsaturated (contains the double C=C bonds) bromine adds across the double C=C and the solution decolorizes (turns colourless). If the molecule being tested does not contain a double C=C bond then the bromine water remains orange.

Worked example

a Give the equation for the reaction between ethene, C_2H_4, and HCl using structural formulae.

b What type of reaction is this?

Answer

a
$$\underset{\underset{H}{|}}{\overset{\overset{H}{|}}{C}} = \underset{\underset{H}{|}}{\overset{\overset{H}{|}}{C}} \quad + \quad H-Cl \longrightarrow H-\underset{\underset{H}{|}}{\overset{\overset{H}{|}}{C}}-\underset{\underset{H}{|}}{\overset{\overset{H}{|}}{C}}-Cl$$

b electrophilic addition

Questions

1 What is meant by heterolytic fission?

2 What is an electrophile?

3 Name the type of reaction that takes place between bromine and ethene.

Unsymmetrical alkenes

Ethene is a **symmetrical** alkene. The groups attached to the C=C bond are the same.

$$
\begin{array}{c}
\text{H} \quad \text{H} \\
| \quad\quad | \\
\text{C} = \text{C} \\
| \quad\quad | \\
\text{H} \quad \text{H}
\end{array}
$$

When symmetrical alkenes such as ethene take part in addition reactions with **electrophiles** such as hydrogen bromide it does not matter how the hydrogen bromide adds across the carbon double bond because the product is always bromoethane.

However, this is not true for all alkene molecules.

Propene is an **unsymmetrical** alkene.

$$
\begin{array}{c}
\text{H} \quad \text{H} \quad \text{H} \\
| \quad\quad | \quad\quad | \\
\text{C} = \text{C} - \text{C} - \text{H} \\
| \quad\quad\quad\quad | \\
\text{H} \quad\quad\quad \text{H}
\end{array}
$$

Notice that the groups attached to the carbon atoms in the C=C bond are different. This is significant when propene takes part in an addition reaction with an electrophile such as HBr. The HBr can add across the bond in two different ways, to produce 2-bromopropane or 1-bromopropane.

The major product for this reaction is 2-bromopropane.

$$
\begin{array}{c}
\text{H} \quad \text{H} \quad \text{H} \\
| \quad\quad | \quad\quad | \\
\text{C} = \text{C} - \text{C} - \text{H} \quad + \quad \text{H} - \text{Br} \\
| \quad\quad\quad\quad | \\
\text{H} \quad\quad\quad \text{H}
\end{array}
$$

$$
\downarrow
$$

$$
\begin{array}{c}
\text{H} \quad \text{H} \quad \text{H} \\
| \quad\quad | \quad\quad | \\
\text{H} - \text{C} - \text{C} - \text{C} - \text{H} \\
| \quad\quad | \quad\quad | \\
\text{H} \quad \text{Br} \quad \text{H}
\end{array}
$$

2-bromopropane

Some 1-bromopropane is also produced.

$$
\begin{array}{c}
\text{H} \quad \text{H} \quad \text{H} \\
| \quad\quad | \quad\quad | \\
\text{C} = \text{C} - \text{C} - \text{H} \quad + \quad \text{H} - \text{Br} \\
| \quad\quad\quad\quad | \\
\text{H} \quad\quad\quad \text{H}
\end{array}
$$

$$
\downarrow
$$

$$
\begin{array}{c}
\text{H} \quad \text{H} \quad \text{H} \\
| \quad\quad | \quad\quad | \\
\text{Br} - \text{C} - \text{C} - \text{C} - \text{H} \\
| \quad\quad | \quad\quad | \\
\text{H} \quad \text{H} \quad \text{H}
\end{array}
$$

1-bromopropane

Polymers

Lots of alkene molecules can be joined together to form **addition polymers**.

- The alkene molecules are called **monomers**.
- This is an addition **polymerization** reaction.

Many ethene molecules can join together to form the addition polymer poly(ethene).

$$
n
\begin{array}{c}
\text{H} \quad \text{H} \\
| \quad\quad | \\
\text{C} = \text{C} \\
| \quad\quad | \\
\text{H} \quad \text{H}
\end{array}
\longrightarrow
\left(
\begin{array}{c}
\text{H} \quad \text{H} \\
| \quad\quad | \\
\text{C} - \text{C} \\
| \quad\quad | \\
\text{H} \quad \text{H}
\end{array}
\right)_n
$$

The formation of poly(ethene).

Notice that poly(alkenes) are **saturated** – they do not have C=C double bonds.

Many propene molecules can join together to form poly(propene).

$$
n
\begin{array}{c}
\text{CH}_3 \quad \text{H} \\
| \quad\quad\quad | \\
\text{C} = \text{C} \\
| \quad\quad\quad | \\
\text{H} \quad\quad \text{H}
\end{array}
\longrightarrow
\left(
\begin{array}{c}
\text{CH}_3 \quad \text{H} \\
| \quad\quad\quad | \\
\text{C} - \text{C} \\
| \quad\quad\quad | \\
\text{H} \quad\quad \text{H}
\end{array}
\right)_n
$$

The formation of poly(propene).

The section drawn in brackets is called a **repeating unit**. The repeating unit is repeated thousands of times in each polymer molecule. The lines representing the covalent bonds between repeating units cross through the brackets around the repeating units.

Worked example

a Draw the structural formula of chloroethene.

b Chloroethene can be polymerized in a similar type of reaction to the polymerization of ethene to poly(ethene). Write the equation for the production of poly(chloroethene) from chloroethene. Clearly show the repeating unit of the polymer.

Answer

a
$$
\begin{array}{c}
Cl \quad H \\
| \quad\; | \\
C = C \\
| \quad\; | \\
H \quad H
\end{array}
$$

b
$$
n\;
\begin{array}{c}
Cl \quad H \\
| \quad\; | \\
C = C \\
| \quad\; | \\
H \quad H
\end{array}
\longrightarrow
\left(
\begin{array}{c}
Cl \quad H \\
| \quad\; | \\
C - C \\
| \quad\; | \\
H \quad H
\end{array}
\right)_n
$$

Why are poly(alkenes) unreactive?

- Alkenes have a sigma and a pi bond.
- The high electron density and easy accessibility of the pi bond means that alkenes can be attacked by electrophiles.
- Poly(alkenes) are saturated hydrocarbons.
  - ® As a result they are much less reactive than alkenes.

Using polymers

Poly(ethene)

- Poly(ethene) is a soft, flexible polymer.
- It is widely used to make plastic bags and bottles.
- Poly(ethene) does not have a sharp melting point.
- It melts over a range of temperatures because it contains a mixture of polymer chains that have different lengths.

Poly(propene)

- Poly(propene) is a tough, strong polymer.
- It is used to make buckets and crates.
- Plastics like poly(propene) can be recycled.
  - First the plastics are separated out.
  - Then the poly(propene) is shredded and then turned into granules.
  - These granules are then melted down and made into useful new items like flower pots.

Worked example

a Poly(tetrafluoroethene) (PTFE) is an expensive polymer. It is formed by an addition polymerization reaction between many tetrafluoroethene molecules. Draw the structural formula of tetrafluoroethene.

b Write the equation for the production of poly(tetrafluoroethene) from tetrafluoroethene. Clearly show the repeating unit of the polymer.

Answer

a
$$
\begin{array}{c}
F \quad F \\
| \quad\; | \\
C = C \\
| \quad\; | \\
F \quad F
\end{array}
$$

b
$$
n\;
\begin{array}{c}
F \quad F \\
| \quad\; | \\
C = C \\
| \quad\; | \\
F \quad F
\end{array}
\longrightarrow
\left(
\begin{array}{c}
F \quad F \\
| \quad\; | \\
C - C \\
| \quad\; | \\
F \quad F
\end{array}
\right)_n
$$

repeating unit

Identifying monomers

To identify the monomer from which an addition polymer has been made, first identify the repeat unit:

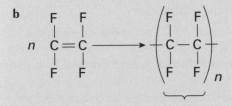

the repeat unit

The monomer which was used to make the polymer must contain the repeat unit with a double C=C bond added and the trailing bonds removed.

$$
\begin{array}{c}
H \qquad\qquad OH \\
\backslash \qquad\quad\; / \\
C = C \\
/ \qquad\quad\; \backslash \\
H \qquad\qquad H
\end{array}
$$

monomer

Questions

1 What is the major product for the reaction between propene and hydrogen bromide?

2 Suggest a use for poly(propene).

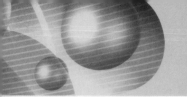

Making margarine

Molecules of vegetable oils contain double C=C bonds. This means most vegetable oils are liquid at room temperature. Many vegetable oils contain many double C=C bonds and are known as **polyunsaturated fats**. However for many uses, such as spreading on bread or making cakes and pastries, a fat which is solid at room temperature is desirable. Vegetable oils can be made into margarine by adding hydrogen across the double C=C bonds in a process known as **hydrogenation**. A nickel catalyst is used. As hydrogen is added to the unsaturated vegetable oil the melting point of the oil increases so that it becomes solid at room temperature.

Dealing with waste polymers

Non-biodegradable plastics do not break down in the environment. Recycling these plastics helps us to protect the environment and reduces the need for their disposal in landfill sites.

First the waste plastics must be separated into different types of plastic. The waste polymers are broken down into small pellets and finally remoulded to make useful new products. Waste products can also be cracked to make monomers which can then be used to make new plastics or other useful chemicals.

Some plastics can be disposed of by incineration. The burning of waste plastics releases heat energy which can be used to generate electricity. PVC contains large amounts of chlorine and the combustion of PVC can produce toxic gases including hydrogen chloride. Chemists minimize damage to the environment by designing apparatus to remove hydrogen chloride from the waste gases from incinerators.

Maize and starch have been used to make a range of biodegradable and compostable polymers. This is very beneficial as the disposal of polymers is a growing problem as more everyday items are being made from plastics.

Biodegradable and compostable polymers are broken down naturally by micro-organisms such as bacteria.

Alcohols

Alcohols contain hydroxyl groups. The presence of these hydroxyl groups means that alcohol molecules can form hydrogen bonds. This affects the physical properties of alcohols.

Boiling points of alcohols

Alcohols have much higher boiling points than alkanes with a similar molecular mass. This is because of the hydrogen bonds which exist between alcohol molecules. Alkane molecules do not form hydrogen bonds.

Alcohol molecules form hydrogen bonds with other alcohol molecules.

A lot of energy is required to overcome the stronger force of attraction between the alcohol molecules due to these hydrogen bonds so alcohols have relatively high boiling points. **Volatility** is a measure of how easy it is for a liquid to turn into a gas. Alcohols have relatively high boiling points, so they have a relatively low volatility.

Solubility in water

Alcohols with short carbon chains are readily soluble in water because the alcohol molecules can form hydrogen bonds with water molecules.

Alcohol molecules form hydrogen bonds with water molecules.

As the length of the carbon chain increases alcohols become less soluble in water.

Uses of alcohols

Methylated sprits is a useful solvent which can be used to remove paint from brushes and clothing. It can also be used as a fuel for camping stoves. Methylated spirits is made by mixing ethanol with methanol. A chemical with a very unpleasant taste is also added together with a purple colouring to warn people not to drink the toxic mixture.

Methanol can be used as a chemical feedstock (starting point) to make many organic chemicals such as plastics.

Combustion of alcohols

The complete combustion of alcohols produces carbon dioxide and water, for example:

$$C_2H_5OH + 3O_2 \rightarrow 2CO_2 + 3H_2O$$

Small amounts of the alcohol methanol and MTBE (an ether made from methanol) are added to unleaded petrol to allow it to burn more easily.

Esterification

Alcohols can be reacted with carboxylic acids to form esters. The reaction is carried out in the presence of an acid catalyst, for example:

ethanol + ethanoic acid ⇌ ethyl ethanoate + water

Esters have pleasant smells and are used in perfumes. They are also used as flavourings and as adhesives and solvents.

The hydration of ethene

- Ethanol can be made by the **hydration** of ethene. In this reaction steam is added to ethene.

Ethene + steam ⇌ ethanol vapour

$$C_2H_4(g) + H_2O(g) \rightleftharpoons C_2H_5OH(g)$$

The reaction is reversible and the forward reaction is exothermic.

- A **phosphoric acid** catalyst is used. This increases the rate of reaction so a dynamic equilibrium is reached more quickly.
- The catalyst does not affect the position of equilibrium so it does not affect the yield of ethanol at equilibrium.

- The forward reaction, between ethene and steam, is exothermic.
- If the system is at equilibrium and the temperature is changed the position of equilibrium will shift to minimize the change in temperature.
- Increasing the temperature increases the rate of reaction but decreases the yield of ethanol (increasing the temperature shifts the position of equilibrium in the endothermic direction).
- Decreasing the temperature increases the yield of ethanol (decreasing the temperature shifts the position in the exothermic direction) but decreases the rate of reaction.

  As a result a compromise temperature of around 300°C is used. This gives a reasonable rate and a reasonable yield of ethanol.

If the system is at equilibrium and the pressure is changed the position of equilibrium will shift to minimize the change in pressure.

- Increasing the pressure shifts the position of equilibrium towards the side with fewer gas molecules so increases the yield of ethanol.
- Decreasing the pressure shifts the position of equilibrium towards the side with more gas molecules decreasing the yield of ethanol.
- However, it is very expensive to maintain a very high pressure. Also at very high pressures the ethene molecules may polymerize to form poly(ethene).

  As a result a pressure of 90 atmospheres is used.

See spread 2.20 for more on equilibria in industrial processes.

Questions

1 Name a toxic gas that may be produced when PVC is burnt.
2 Why do alcohols have higher boiling points than alkenes with similar molecular masses?
3 Suggest how esters can be used.

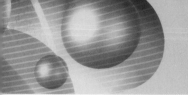

Fermentation

- Ethanol is a type of **alcohol**.

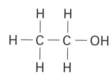

- Alcohols have the general formula $C_nH_{2n+1}OH$.
- Alcohols contain the OH or **hydroxyl** group.
- Alcohols that have more than two carbon atoms can exist as structural isomers, for example propan-1-ol and propan-2-ol. The number indicates the position of the hydroxyl group.

propan-1-ol   propan-2-ol

Using ethanol and methanol as fuels

The alcohols methanol and ethanol are both liquids at room temperature. Methanol and ethanol are both useful fuels. Liquid fuels are much easier to handle and store than gaseous fuels such as hydrogen and methane:

- Liquid fuels are easier to transfer than gases.
- Liquids fuels have a smaller volume than gases so smaller fuel tanks can be used.
- Also gaseous fuels would need to be kept under high pressures, so they are more likely to escape than liquids.

Ethanol made by the **fermentation** of plant material such as sugar cane is a **renewable fuel**.

- Ethanol can also be made by the hydration of ethene. As the ethene is obtained from crude oil this is non-renewable.
- Ethanol and methanol both burn very cleanly, releasing very little carbon monoxide.
- Methanol is normally produced from methane, which is obtained from natural gas so is non-renewable.
- Methanol can also be made from the reaction between hydrogen and carbon monoxide.

Worked example

Give two reasons why motorists may prefer liquid fuels such as ethanol to gaseous fuels such as hydrogen.

Answer

Liquid fuels are easier to transfer to the vehicle and because liquids take up a smaller volume than gases smaller fuel tanks can be used.

Industrial alcohol

- Fermentation can also be used to make ethanol for use as fuels or for industrial purposes.
- The sugar comes from a wide range of sources including sugar beet or sugar cane.
- The reaction is carried out at around 37°C in aqueous anaerobic conditions.

Biofuels

- Biofuels such as wood are fuels that are made from living things.
- Ethanol produced by the fermentation of sugar is a **biofuel**.
- A **carbon-neutral** fuel releases no net carbon emissions to the atmosphere over a year.

Ethanol produced by the **hydration** of ethene is not carbon-neutral because the ethene comes from crude oil and when the ethanol is burnt the carbon is released into the atmosphere as carbon dioxide.

Ethanol produced by the fermentation of corn is also not carbon-neutral. Farmers use large amounts of fossil fuels

- to make the fertilizers used to increases the yields of corn
- to fuel the tractors to plant and harvest the corn

Some carbon dioxide is absorbed from the atmosphere as the corn grows but this carbon is released into the atmosphere when the ethanol is burnt.

However, ethanol produced from sugar cane can be carbon-neutral.

As the sugar cane grows it absorbs carbon dioxide from the atmosphere. The sugar cane is then converted into ethanol. Any waste sugar cane can be used to fuel the process, for example as a fuel for vehicles. When the alcohol is burnt the carbon dioxide is released back into the atmosphere. This could be very important in the future.

Ethanol can be made industrially by the hydration of ethene

$$C_2H_4 + H_2O \rightarrow C_2H_5OH$$

This is an addition reaction. Addition reactions have an **atom economy** of 100%. This is good for sustainable development. All the reactant atoms are made into useful products.

Ethanol can also be made industrially by the fermentation of glucose from plants.

$$Glucose \rightarrow ethanol + carbon\ dioxide$$

$$C_6H_{12}O_6 \rightarrow 2CH_3CH_2OH + 2CO_2$$

This is a decomposition reaction. Carbon dioxide is also produced by the reaction although this can be used in the manufacture of fizzy drinks.

Wine and beer

- Fermentation on a large scale is used to produce wine and beer.
- The sugar in barley is used to make beer.
- The sugar in grape juice is used to make wine.

Fermentation versus hydration of ethene

fermentation	hydration of ethene
slower rate of reaction.	faster rate of reaction.
yield of ethanol of around 15%.	yield of ethanol of around 95%.
batch process. It is more labour intensive, so labour costs are higher but set up costs are lower.	**continuous process**, so labour costs are less but set-up costs are higher.
atom economy of 51.1%.	an addition reaction with an atom economy of 100%
uses lower temperatures and pressures so uses lower amounts of energy.	uses higher temperatures and pressures so uses higher amounts of energy.
renewable	non-renewable
Distillation is used to increase the concentration of alcohol. The alcohol that is produced has a lower purity so more steps are required for purification.	Distillation is used to increase the concentration of alcohol. The alcohol produced is already purer so purification is easier.

Worked example

Ethanol can be made by fermentation or by the hydration of ethene.

The ethanol produced is then purified.

a How is the ethanol purified?

b Why is the purification of ethanol made by the hydration of ethene easier than the purification of ethanol made by fermentation?

Answer

a Distillation

b Ethanol made by the hydration of ethene is already purer.

Green chemistry

Ethanol produced by fermentation from sugar cane or sugar beet is a renewable resource. Sometimes there are unforeseen and undesirable consequences to green policies. If large areas of land are used to grow crops to produce ethanol rather than to grow food for people to eat the benefit of using a renewable energy resource rather than a non-renewable fossil-fuel resource will be less important than local people having enough food to eat.

Chemical sustainability

Chemists are very aware of the principles of chemical sustainability to allow for a good standard of living today and in the future. Chemists try and choose industrial processes which do not involve the use of harmful chemicals and reduce the overall number of chemicals used. They also try and use alternative energy resources such as wind and solar rather than using non-renewable fossil fuels. Finally chemists try and ensure that the waste products made in chemical processes are non-toxic and can either be recycled into useful new products or biodegraded into harmless substances in the environment.

Types of alcohol

We can classify alcohols as primary, secondary, or tertiary depending on the number carbon atoms attached to the carbon atom bonded to the hydroxyl, OH, group.

Tertiary alcohols

In tertiary alcohols the carbon atom attached to the hydroxyl group is attached to three other carbon atoms.

Secondary alcohols

In secondary alcohols the carbon atom attached to the hydroxyl group is attached to two other carbon atoms.

Primary alcohols

In primary alcohols the carbon atom attached to the hydroxyl group is attached to one other carbon atom.

A tertiary alcohol   A secondary alcohol   A primary alcohol

Worked example

Draw the structural formula of the tertiary alcohol with the formula $C_4H_{10}O$.

Answer

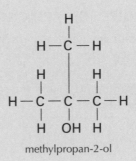

methylpropan-2-ol

Questions

1 Give two ways that ethanol can be made.

2 What is a carbon-neutral fuel?

3 What type of alcohol is propan-2-ol?

Tertiary alcohols

Tertiary alcohols such as methylpropan-2-ol have a hydroxyl group attached to a carbon atom that is attached to three other carbon atoms.

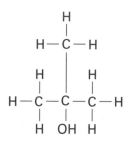

- Tertiary alcohols are not oxidized by oxidizing agents such as aqueous acidified potassium dichromate(VI), $K_2Cr_2O_7$.
  - As a result if acidified potassium dichromate(VI) solution is added to a tertiary alcohol no oxidation reaction takes place so the orange solution stays orange.

Primary and secondary alcohols

Primary and secondary alcohols are oxidized by oxidizing agents such as aqueous acidified potassium dichromate(VI), $K_2Cr_2O_7$.

Primary alcohols

Primary alcohols are first oxidized to **aldehydes**.

$$CH_3CH_2OH + [O] \rightarrow CH_3CHO + H_2O$$

Notice how the oxidizing agent is written as [O] but it is important that the equation still balances.

As the primary alcohols is oxidized the chromium(VI) ion is reduce to chromium(III) and the solution changes colour from orange to green.

If the oxidizing agents are in excess then the aldehyde is oxidized to a **carboxylic acid**.

$$CH_3CHO + [O] \rightarrow CH_3COOH$$

As the aldehyde is oxidized the chromium(VI) ion is reduced to chromium(III) and the solution changes colour from orange to green.

Secondary alcohols

Secondary alcohols are oxidized to **ketones** but they cannot be oxidized any further.

$$CH_3CH(OH)CH_3 + [O] \rightarrow CH_3COCH_3 + H_2O$$

Elimination of water to form alkenes

Dehydrating agents such as concentrated sulfuric acid or phosphoric acid can be used to remove H_2O from alcohols to form alkenes. The reaction is carried out at a high temperature.

Example

propan-1-ol → propene + water

$$CH_3CH_2CH_2OH \rightarrow CH_3CHCH_2 + H_2O$$

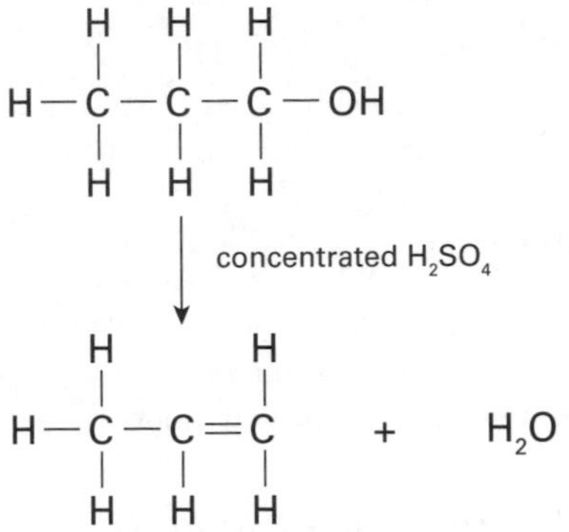

concentrated H_2SO_4

Notice that as water is lost from the alcohol during the reaction this may also be classified as an **elimination** reaction.

Producing polymers from alcohols

Alkenes made by the elimination of water from alcohols could be used as raw materials to make **polymers**.

Here the **monomer** propene is used to make the polymer poly(propene).

Polymers produced from alcohols made by the fermentation of plant materials provide an alternative to using monomers derived from crude oil, which is a non-renewable resource.

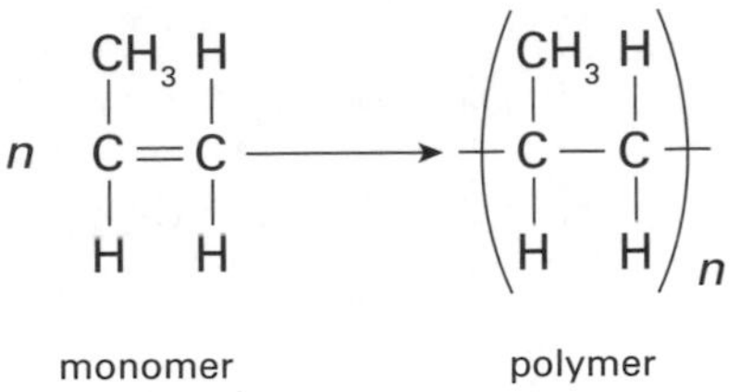

monomer   polymer

Questions

1 What is formed when a secondary alcohol is oxidized?

2 Give two reagents that could be used to differentiate between an aldehyde and a ketone.

3 Suggest a dehydrating agent that could be used to form an alkene from an alcohol.

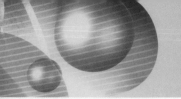

2.14 Halogenoalkanes

More about halogenoalkanes

The halogenoalkanes, $C_nH_{2n+1}X$, are also known as **haloalkanes** and **alkyl halides**.

They are alkanes in which one hydrogen atom has been substituted by a halogen atom, X, where X can be fluorine, chlorine, bromine, or iodine.

Examples of halogenoalkanes:

1-bromopropane

2-bromopropane

2-bromo-2-methylpropane

Polar bonds

Most halogen atoms are more electronegative than carbon. This means that haloalkanes have polar C–X bonds.

(Notice that X is used to represent a halogen atom.)

Nucleophiles

Nucleophiles such as OH⁻, CN⁻, and NH₃ have a lone pair of electrons which they can donate to another molecule to form a new covalent bond.

The nucleophilic substitution of primary halogenoalkanes

Hydroxide ions

Halogenoalkanes undergo **nucleophilic substitution** reactions with hydroxide ions to form alcohols.

This reaction can also be described as **hydrolysis**.

A hot aqueous alkali is used.

Strength of the C–X bond

The rate of reaction of different halogenoalkenes can be compared by using either:

- aqueous silver nitrate in ethanol (the water acts as the nucleophile)

or

- hot aqueous alkali followed by neutralization and then aqueous silver nitrate (the hydroxide ions act as the nucleophile).

The reactivity of halogenoalkanes depends on the halogen involved.

- **Polar bonds** are more **susceptible** to attack from nucleophiles than non-polar bonds.
- Iodoalkanes react faster with hydroxide ions than bromoalkanes or chloroalkanes.

The C–Cl bond has a very high bond enthalpy. This means that the bond is very strong and hard to break. The **activation energy** required to break the bond and start the reaction is so high that the reaction is very slow.

bond	bond enthalpy (kJ mol^{-1})
C–Cl	338
C–Br	276
C–I	238

- Remember that the weaker the bond the faster the hydrolysis reaction.
  - As a result **bond enthalpy** is a more important factor than polarity.

The importance of conditions used

The reaction between halogenoalkanes and hydroxide ions can produce different products depending on the conditions chosen. Substitution occurs in aqueous conditions.

Substitution

If the reaction is carried out using water as a solvent

The hydroxide ion acts as a nucleophile.

Mechanism for the substitution reaction

Using nucleophilic substitution reactions

- The nucleophilic substitution reactions of halogenoalkanes are used to introduce new functional groups to organic molecules.

Questions

1 What is a nucleophile?

2 Why do iodoalkanes react faster than chloroalkanes?

3 What is the organic product of the reaction between 2-bromopropane and hydroxide ions in aqueous conditions?

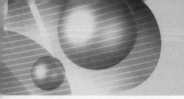
Modern instrumental methods

The **mass spectrometer** and **infrared spectrometer** can be used to identify compounds.

Compared with traditional laboratory techniques, these modern methods of analysis are:

- faster
- more accurate
- more sensitive
- able to use smaller samples

These spectrometers can be connected directly to computers which can process enormous amounts of information at very high speed. Modern instrumental methods can also identify how much of the compound is present.

Mass spectroscopy

Key definition

The relative isotopic mass of an atom of an isotope of an element is measured on a scale where an atom of carbon-12 is 12.

Mass spectrometry can be used to determine relative isotopic masses.

The mass spectrum of this lead sample shows it contains four isotopes. The relative isotopic masses of these isotopes of lead are 204, 206, 207, and 208. The relative abundance of each isotope can be read from the mass spectrum or may be given in a table.

The mass spectrum of lead

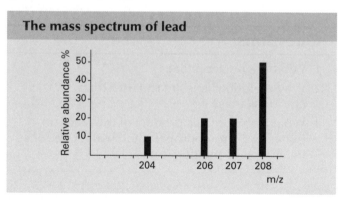

Relative atomic mass, A_r

The mass of all atoms is measured relative to the mass of a ^{12}C atom. Relative atomic mass can be calculated using the data from a mass spectrum. From the mass spectrum of lead above we obtain the following data:

m/z	204	206	207	208
relative abundance (%)	10	20	20	50

- m/z is the same as the mass of the ion if the charge is +1.
- relative abundance tells us the proportion of each isotope present in the sample.
- the A_r of lead can then be calculated as a weighted mean average.

$$A_r = \frac{(204 \times 10) + (206 \times 20) + (207 \times 20) + (208 \times 50)}{100} = 207$$

When organic molecules pass through a mass spectrum they fragment (break up).

By studying the position of the peaks caused by fragmentation chemists can suggest the identity of the ions responsible. This allows chemists to build up an idea of the structure of the organic compound being investigated.

Mass/charge	Ion responsible for the peak
15	CH_3^+
29	$C_2H_5^+$
43	$C_3H_7^+$
17	OH^+

Worked example

The mass spectrum of an alcohol shows peaks at:

- 15 m/z
- 17 m/z
- 29 m/z
- 31 m/z
- 46 m/z

a What is the M_r of the compound?

b Suggest the identity of the ion responsible for each of the peaks observed.

c Suggest the structure of this compound.

Answer

a The peak with the highest mass/charge ratio has a value of 46 m/z, so the M_r of the compound is 46.

b 15 m/z = CH_3^+

17 m/z = OH^+

29 m/z = $C_2H_5^+$

31 m/z = CH_2OH^+

46 m/z = $C_2H_5OH^+$

c
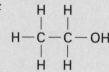

Even if two compounds have the same M_r and the same atoms, if they have different structures they will fragment in different ways.

Mass spectrometry can be used to monitor the levels of pollutants such as lead in the environment. It has also been used by the space probes sent to Mars to identify the elements present on the surface of the planet.

Infrared spectroscopy

Infrared spectroscopy is used to identify organic molecules. The covalent bonds between atoms are like springs. In infrared spectroscopy infrared radiation is used to make the bonds vibrate.

Different types of bonds absorb infrared radiation of slightly different wavelength. By seeing which wavelengths have been absorbed we can identify **functional groups** in organic molecules.

typical wavenumber (cm^{-1})	bond
750–1100	C–C
1000–1300	C–O
1620–1680	C=C
1640–1750	C=O
2500–3300	O–H in carboxylic acids
2850–3100	C–H
3200–3550 (broad)	O–H bond in alcohols, phenols
3200–3500	N–H

We measure the infrared radiation absorbed using wavenumbers. The wavenumber is the number of waves in 1 cm. The higher the frequency the more waves in 1 cm so the higher the wavenumber.

Infrared spectroscopy is used to monitor the levels of air pollution in the atmosphere. This technique can be used to monitor the levels of pollutants such as CH_4, NO, and SO_2. The covalent bonds in these molecules absorb radiation of different wavelengths and this can be used to measure how much of the polluting gas is present in the air sample.

Modern breathalysers use infrared spectroscopy to measure the amount of ethanol in the breath of drivers suspected of having been drinking alcohol.

Infrared spectrum of ethanol

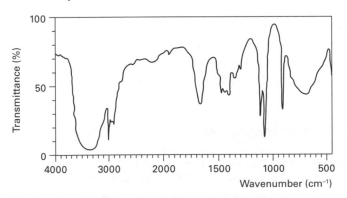

Notice that the infrared spectrum of ethanol shows a broad peak at 3400 cm^{-1} caused by the O–H bond and another at 1100 cm^{-1} caused by the C–O bond.

Useful ideas

Alkanes

Alkanes will contain C–H bonds so they will have an absorption band between 2850 and 3300 cm^{-1} and probably also C–C bonds so they will also have an absorption band between 750 and 1100 cm^{-1}

Alkenes

Alkenes will also contain C–H bonds and will also probably contain C–C bonds. However, they will also contain C=C bonds so they will have an absorption band between 1620 and 1680 cm^{-1}.

Carbonyls

Aldehydes and ketones both contain C=O bonds so they will have an absorption band between 1640 and 1750 cm^{-1}.

Carboxylic acids

Carboxylic acids also contain C=O bonds so they will have an absorption band between 1640 and 1750 cm^{-1}. In addition they contain an O–H bond so they will have an absorption band between 2500 and 3300 cm^{-1}.

Alcohols

Alcohols contain a C–O bond so they will have an absorption band between 1000 and 1300 cm^{-1}
They will also have a O–H bond so they will have a broad absorption band between 3200 and 3550 cm^{-1}

Ethers

Ethers contain C–O bonds so they will have an absorption band between 1000 and 1300 cm^{-1}.

Impurities

Any impurities in a sample will produce absorption bands that should not be there.

The fingerprint region

The area of an infrared spectrum between 400 cm^{-1} and 1500 cm^{-1} is known as the **fingerprint region** because it is unique for every compound.

An unknown sample can be identified by comparing its infrared spectrum with a database of known infrared spectra to find a match.

Global warming

Carbon dioxide, methane, and water vapour are greenhouse gases. The bonds in these molecules absorb infrared radiation from the Earth's surface. The radiation is then emitted in all directions. Although some of this radiation is lost into space some increases the temperature of the atmosphere, resulting in global warming.

Questions

1 What is special about high-resolution mass spectroscopy?
2 Name three greenhouse gases.
3 How can you use infrared spectroscopy to show that a sample contains impurities?

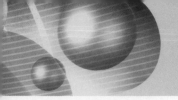

- Enthalpy, *H*, is the heat energy stored in a chemical system.

- Enthalpy change, ΔH, is the heat energy change at constant pressure.

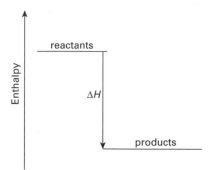

The enthalpy diagram for an exothermic reaction

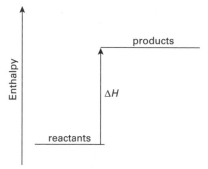

The enthalpy diagram for an endothermic reaction

Standard enthalpy change of formation

- The **standard enthalpy of formation**, $\Delta H_f^{\ominus}$ is the enthalpy change when one mole of a compound is formed from its elements in their standard states, under standard conditions.

  - As a result the standard enthalpy change of formation of an element in its standard state is zero

Exothermic and endothermic reactions

Energy must be taken in to break existing bonds, while energy is given out when new bonds are formed. During chemical reactions there is often a difference between the amount of energy taken in to break the existing bonds and the amount of energy given out when the new bonds are formed. Most reactions give out energy overall. These reactions are described as **exothermic**.

Exothermic reactions

The combustion of fuels and the oxidation of carbohydrates are examples of exothermic reactions.

- The substances involved in exothermic reactions get hotter because chemical energy is being changed into thermal (heat) energy.
- During the reaction the chemicals lose energy. The energy that is lost by the chemicals is gained by the surroundings.
  - As a result the enthalpy change (ΔH) for an exothermic reaction is always negative.

The more exothermic a reaction is the more likely it is to happen, although other factors such as the activation energy will determine whether the reaction will actually take place.

Endothermic reactions

Some reactions take in more energy to break existing bonds then they release when new bonds are formed. These reactions are described as **endothermic**. The thermal decomposition of calcium carbonate is an example of an endothermic reaction.

- The substances involved in endothermic reactions get colder because thermal energy is taken from the surroundings.
- The energy that is lost by the surroundings is gained by the chemicals.
  - As a result the enthalpy change (ΔH) for an endothermic reaction is always positive.

Standard enthalpy changes

The amount of energy given out or taken in depends on the temperature and pressure.

We use standard enthalpy changes so that different enthalpy changes can be compared. This means that all the other **conditions** must be kept constant.

- 1 mole of a substance is reacted.
- Any solutions have a concentration of 1.00 mol dm^{-3}.
- Any gases must have a pressure of 100 kPa.
- A stated temperature must be used, normally 25°C.
- Elements must be in their standard states, for example carbon as graphite not as diamond.

Standard enthalpy change of combustion

The **standard enthalpy of combustion**, $\Delta H_c^{\ominus}$, is the enthalpy change when one mole of a compound is completely burnt in oxygen under standard conditions.

The reaction involving $\Delta H_c^{\ominus}$ of methane, CH_4, is represented as

$$CH_4(g) + 2O_2(g) \rightarrow CO_2(g) + 2H_2O(g)$$

$$\Delta H_c^{\ominus} \text{ at } 25°C = -890.3 \text{ kJ mol}^{-1}$$

Calorimetry

Calculating the heat change of a reaction

The heat energy change for a reaction (Q, measured in joules) is given by the equation

$$Q = m \, c \, \Delta T$$

Where m = the mass of the surroundings (g), c = the specific heat capacity of the surroundings, (J g^{-1} °C^{-1}) and ΔT is the temperature change (final temperature – initial temperature), (°C).

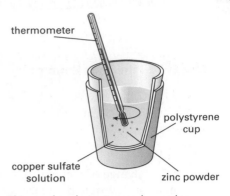

thermometer

polystyrene cup

copper sulfate solution

zinc powder

This simple calorimeter can be used to calculate the heat energy change during the displacement reaction between zinc and copper sulfate.

Calculating the enthalpy change of a reaction

- Enthalpy changes are measured in kJ mol^{-1}

- Divide the heat change for the reaction by 1000

- Then divide the heat energy change by the number of moles involved.

Calorimetry

Calorimetry can be used to calculate the heat energy change of a reaction.

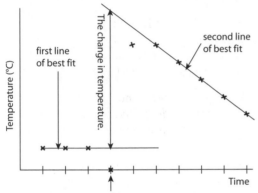

first line of best fit

second line of best fit

The change in temperature.

Temperature (°C)

Time

at X Zinc is added to the Copper Sulfate Solution

Measuring the change in temperature.

Worked example

In an experiment an excess of zinc powder was added to 50 cm^3 of 1.00 mol dm^{-3} copper sulfate solution.

The initial temperature was 20.0°C.

The final temperature was 31.0°C.

Find the enthalpy change for the reaction:

$Zn(s) + CuSO_4(aq) \rightarrow Cu(s) + ZnSO_4(aq)$

Copper sulfate solution has a density of 1.00 g cm^{-3} and a specific heat capacity of 4.18 J g^{-1}°C^{-1}

The energy change for the reaction = $mc\Delta T$

The mass of 50 cm^3 of copper sulfate solution = 50 g

The specific heat capacity of the copper sulfate solution is 4.18 J g^{-1}°C^{-1}, so the temperature change
= 31.0 − 20.0 = 11.0°C

The heat gained by the surroundings
= 50 g × 4.18 J g^{-1}°C^{-1} × 11°C = 2299 J

The heat lost by the chemical system = 2299 J or 2.299 kJ

The amount of copper sulfate solution that reacted in this experiment:

Number of moles = (volume/1000) × concentration

= (50/1000) × 1.00 = 0.05 moles

The enthalpy change in kJ mol^{-1}
= 2.299 kJ /0.05 moles = 45.98 kJ mol^{-1}

$Zn(s) + CuSO_4(aq) \rightarrow Cu(s) + ZnSO_4(aq)$
$\Delta H = 45.98$ kJ mol^{-1}

Questions

1 What happens to the thermal energy lost by the chemicals in an exothermic reaction?

2 Why do endothermic reactions get colder?

3 What is $\Delta H_f^{\ominus}$ of $N_2(g)$?

First law of thermodynamics

Energy cannot be created or destroyed; it is merely changed from one form to another.

Hess's law

The enthalpy change of a chemical reaction is **independent** of the route by which the reaction is achieved and depends only on the initial and final states.

🔊 As a result you can use **Hess's law** to find enthalpy changes for reactions that cannot be measured directly in the laboratory.

- The enthalpy change from the reactants A to the products B is the same as the sum of the enthalpy changes from A to C and from C to B.
- We can use $\Delta H_f^{\ominus}$ and $\Delta H_c^{\ominus}$.

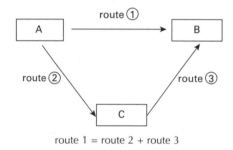

route 1 = route 2 + route 3

Using enthalpies of combustion

The standard enthalpy of combustion, $(\Delta H_c^{\ominus})$ is the enthalpy change when one mole of a compound is completely burnt in oxygen under standard conditions.

- Notice that the arrow is drawn from the chemicals to the combustion products.

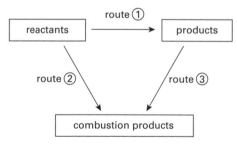

Using enthalpy of combustion values: route 1 = route 2 − route 3

Worked example

Calculate the enthalpy change for

$$C(s) + 2H_2(g) \rightarrow CH_4(g)$$

Given that

$\Delta H_c^{\ominus}$ C(s) $= -394$ kJ mol^{-1}

$\Delta H_c^{\ominus}$ H$_2$(g) $= -286$ kJ mol^{-1}

$\Delta H_c^{\ominus}$ CH$_4$(g) $= -890$ kJ mol^{-1}

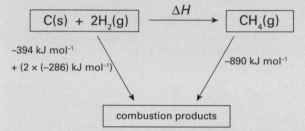

The Hess's law diagram for the formation of methane.

The unknown enthalpy change $\Delta H^{\ominus}$

$= -394$ kJ mol^{-1} + (2 × −286 kJ mol^{-1}) − (−890 kJ mol^{-1})

$= -76$ kJ mol^{-1}

Worked example

Calculate the enthalpy change for the hydrogenation of ethene to form ethane.

$$C_2H_4(g) + H_2(g) \rightarrow C_2H_6(g)$$

Given that

$\Delta H_c^{\ominus}$ C$_2$H$_4$(g) $= -1409$ kJ mol^{-1}

$\Delta H_c^{\ominus}$ H$_2$(g) $= -286$ kJ mol^{-1}

$\Delta H_c^{\ominus}$ C$_2$H$_6$(g) $= -1560$ kJ mol^{-1}

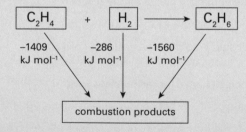

The Hess's law diagram for the hydrogenation of ethene.

The unknown enthalpy change $\Delta H^{\ominus}$

$= -1409$ kJ mol^{-1} −286 kJ mol^{-1} − (−1560 kJ mol^{-1})

$= -135$ kJ mol^{-1}

Using enthalpies of formation

The standard enthalpy of formation, $\Delta H_f^\ominus$ is the enthalpy change when one mole of a compound is formed from its elements in their standard states under standard conditions.

- Notice that the arrow is drawn from the elements to the chemicals.

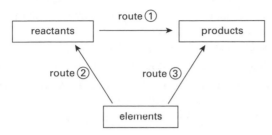

Using standard enthalpy of formation values:
route 1 = − route 2 + route 3

Worked example

Calculate the enthalpy change for the combustion of ethene, $C_2H_4(g)$.

$$C_2H_4(g) + 3O_2(g) \rightarrow 2CO_2(g) + 2H_2O(g)$$

Given that

$\Delta H_f^\ominus\ C_2H_4(g)\ =\ +52\ kJ\ mol^{-1}$

$\Delta H_f^\ominus\ CO_2(g)\ =\ -394\ kJ\ mol^{-1}$

$\Delta H_f^\ominus\ H_2O(g)\ =\ -286\ kJ\ mol^{-1}$

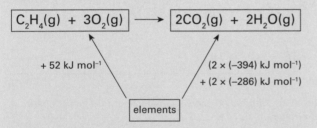

The Hess's law diagram for the combustion of ethene.

The unknown enthalpy change $\Delta H^\ominus$

$= -\ (+\ 52\ kJ\ mol^{-1}) + (2 \times -394\ kJ\ mol^{-1})$
$\quad + (\ 2 \times -\ 286\ kJ\ mol^{-1})$

$= -1412\ kJ\ mol^{-1}$

Mean bond enthalpies

The **mean (average) bond enthalpy** is the mean amount of energy required to break one mole of a specified type of covalent bond in a gaseous species.

$$A-B(g) \rightarrow A(g) + B(g)$$

- This average is obtained from the values obtained from many molecules so the actual values will probably be a little different.
- Energy is required to break bonds.
- Different bonds may require different amounts of energy

$$CH_4(g) \rightarrow C(g) + 4H(g)$$

$$\Delta H = +\ 1664\ kJ\ mol^{-1}$$

The average bond enthalpy for a C–H bond

$$= \frac{1664\ kJ\ mol^{-1}}{4} = 416\ kJ\ mol^{-1}$$

- Bond enthalpies are always endothermic (positive).
- Energy is released when new bonds are formed.

Worked example

Calculate the enthalpy change when methane, $CH_4(g)$ is burnt.

bond	average bond enthalpy (kJ mol^{-1})
C–H	413
O=O	498
C=O	805
O–H	464

$$CH_4(g) + 2O_2(g) \rightarrow CO_2(g) + 2H_2O(g)$$

Energy required to break the existing bonds

$4 \times C–H = 4 \times 413\ kJ\ mol^{-1} = 1652\ kJ\ mol^{-1}$

$2 \times O=O = 2 \times 498\ kJ\ mol^{-1} = 996\ kJ\ mol^{-1}$

Total = 2648 kJ mol^{-1}

Energy released when the new bonds are formed

$2 \times C=O = 2 \times 805\ kJ\ mol^{-1} = 1610\ kJ\ mol^{-1}$

$4 \times O–H = 4 \times 464\ kJ\ mol^{-1} = 1856\ kJ\ mol^{-1}$

Total = 3466 kJ mol^{-1}

The net enthalpy change = 2648 kJ mol^{-1} − 3466 kJ mol^{-1} = −818 kJ mol^{-1}

Questions

1 What is Hess's law?

2 Why do we use Hess's law?

3 What is mean bond enthalpy?

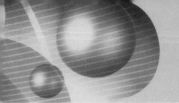

2.18 Collision theory and Boltzmann distributions

Collision theory

You can use the **collision theory** to understand how the conditions used affect the rate of a chemical reaction.

- For a reaction to occur particles must collide.

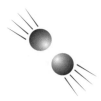

Particles must collide before they can react.

- When the particles collide they must have enough energy to break the existing bonds.
- When the particles collide they must collide in the correct **orientation** so that the reactive parts of the molecules come together. This is particularly important for large molecules.
  - Ⓡ As a result only a small proportion of collisions between particles result in a reaction.

Activation energy

Activation energy, E_a, is the minimum collision energy that particles must have to react.

- E_a is the minimum amount of energy that the reactants must have to form an **activated complex** in a transition state. In an activated complex the old bonds are partially broken and the new bonds are partially made.
- Different reactions have different activation energies.
- The lower the activation energy the larger the number of particles that can react at any temperature.

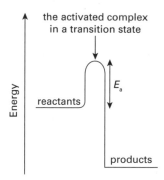

The energy change in an exothermic reaction

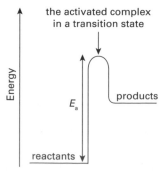

The energy change in an endothermic reaction

The effect of changing the temperature

- As the temperature of a sample of particles changes the kinetic energy of the particles changes.
- As you increase the temperature of a sample you increase the kinetic energy of the particles.
- As the temperature increases, the particles move faster so they collide more often.
- Also when the particles do collide more of the particles will have enough energy to react.
  - Ⓡ As a result the higher the temperature the larger the number of particles that can react.

However, the particles will only react if they collide in the correct orientation.

Boltzmann distributions

These diagrams are used to represent the energy of the particles in a sample of a gas at a given temperature.

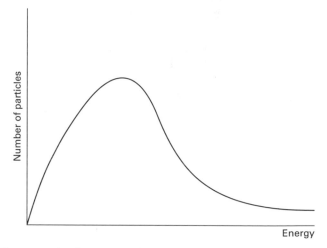

The energy distribution curve for a sample of gas at a particular temperature.

- Notice that the **distribution curve** is not symmetrical.
- Most of the particles have an energy which falls within quite a narrow range, with few particles having much more or much less energy.
- The line does not cross the *x*-axis at higher energy. In fact it would only do so at infinity.
- The line starts at the origin, showing that none of the particles has no energy.
- The total area under the distribution curve represents the total number of gas particles.

Activation energy and Boltzmann distributions

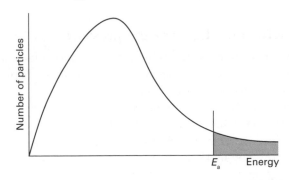

The activation energy is the minimum collision energy that particles must have to react.

- For a reaction to occur the particles must collide and these collisions must have energy equal to or greater than the activation energy. If there is not enough energy when the particles collide then the particles will not react.
- Only the small proportion of particles in the shaded part of the distribution curve have enough energy to react.
- Notice that the activation energy is drawn towards the right of the peak of the distribution curve. If more than half the particles had enough energy to react, the reaction would be too fast to be controlled safely.

Boltzmann distributions and temperature

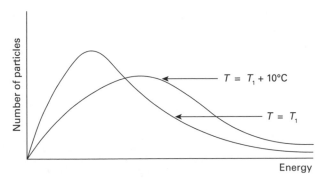

The Boltzmann distributions at different temperatures.

Notice how the range of energies that the particles have increases as the temperature increases.

- The average energy of the particles increases.
- The peak of the distribution curve moves to the right as the average energy increases.
- The distribution curve becomes flatter because the total number of particles remains the same.

Worked example

The graph below shows the Boltzmann distribution for a sample at a particular temperature T_1.

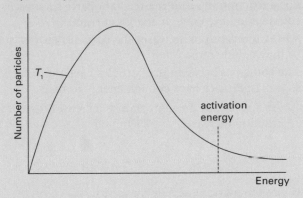

a Shade the proportion of particles that have enough energy to react at temperature T_1.

b Sketch a curve to show the Boltzmann distribution of energies at temperature T_2, which is 10°C hotter than T_1.

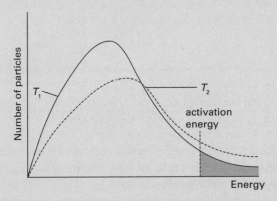

- Notice that only the particles that have energy greater than the activation energy have enough energy to react.

- Notice how at the higher temperature T_2, the average energy of the particles has increased so the peak of the graph has moved to the right and the curve has become flatter.

Questions

1 What is the activation energy of a reaction?
2 How does increasing the temperature affect the energy of the particles?
3 What do Boltzmann diagrams show?

2.19 The effect of temperature, concentration, and catalysts

The effect of changing the temperature

- For a reaction to occur the reactant particles must collide and have enough energy to react.
- As the temperature increases the particles move faster so they collide more often.
- The collisions between the particles have more energy, so more particles have enough energy to react.
  - Ⓡ As a result, as the temperature increases the rate of reaction increases.

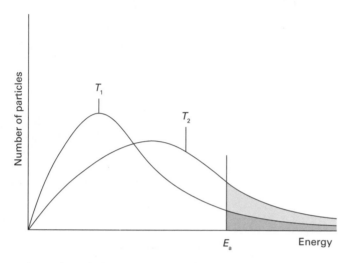

The Boltzmann distributions at temperature T_1 and the higher temperature T_2.

- Notice that as the temperature increases a much higher proportion of the particles have energy greater than or equal to the activation energy.
  - Ⓡ As a result even a modest increase in temperature can lead to a very large increase in the rate of a chemical reaction.

The effect of changing the concentration

- For a reaction to occur the particles must collide.
- As the concentration of a solution increases the distance between the reactant particles decreases.
- The particles have to travel less distance before colliding so they collide more often.
  - Ⓡ As a result as the concentration increases the rate of reaction increases.

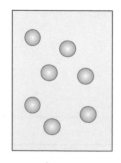
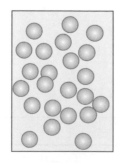

| lower concentration | higher concentration |

Increasing the concentration decreases the distance between the particles.

The effect of changing the pressure

- For a reaction to occur the particles must collide.
- As the pressure of a gas increases the distance between the reactant particles decreases.
- The particles have to travel less distance before colliding so they collide more often.
  - Ⓡ As a result as the pressure increases the rate of reaction increases.

Catalysts

- A catalyst increases the rate of a chemical reaction but is not used up itself during the reaction.
- Catalysts work by providing an alternative reaction pathway which has **lower activation energy**.
  - Ⓡ As a result at any given temperature more particles will have energy greater than or equal to the activation energy.
  - Ⓡ As a result the reactions happen faster (the rate of reaction increases).

Catalysts and Boltzmann distributions

- Adding a catalyst does not change the distribution curves but it does mean that more particles have enough energy to react.

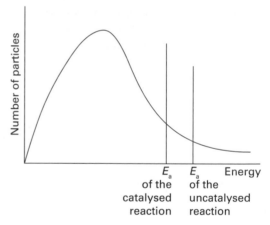

A catalysed reaction has a lower activation energy.

Catalysts and energy level diagrams

Uncatalysed reactions

- In an uncatalysed reaction the activation energy is the minimum amount of energy that must be supplied to form an activated complex in a **transition state**.

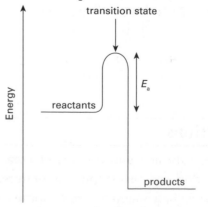

The activation energy in an uncatalysed exothermic reaction.

80

Catalysed reactions

- In catalysed reactions the reactants form an **intermediate**.
- The intermediate then reacts to form a product.
- The activation energy for a catalysed reaction is lower than that for an uncatalysed reaction.
  - As a result a catalyst increases the rate of a reaction.

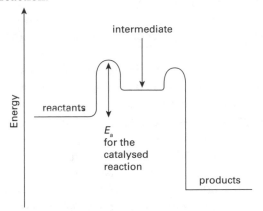

The activation energy in a catalysed exothermic reaction.

Why use catalysts?

Catalysts are used in industry because they affect the conditions that are required to carry out reactions. A suitable catalyst may allow a reaction to be carried out at a lower temperature and pressure than would be otherwise suitable. This means that less energy is required to achieve and maintain the required reaction temperatures and pressures so less fossil fuel needs to be burnt. This means less carbon dioxide is produced.

Enzymes

Enzymes (biological catalysts) can be used to catalyse some reactions. Enzymes are protein molecules and many enzymes work well at room temperatures and pressures. This is very useful in industry as these relatively low temperatures and pressures can be achieved without the need for burning lots of fossil fuels. Burning fossil fuels is undesirable because it is economically very expensive and also results in the production of large amounts of carbon dioxide.

Useful catalysts

- Iron is used as a catalyst in the Haber process which is used to produce ammonia.
- A Ziegler-Natta catalyst is used to catalyse the production of polythene. Polythene made using this catalyst is denser, more rigid, and has a higher melting point.
- Alloys of platinum, palladium, and rhodium are used in the catalytic converters fitted to cars.

Worked example

The graph below represents the Maxwell–Boltzmann distribution of energies at a particular temperature.

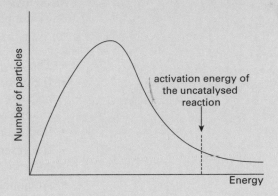

State the effect of adding a catalyst on:

a the energy of the particles

b the activation energy

c the rate of reaction

Answer

a Adding a catalyst will not affect the energy of the particles.

b Adding a catalyst will provide an alternative reaction pathway with a lower activation energy

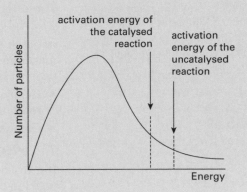

c The rate of reaction will increase because more particles now have enough energy to react.

Questions

1 Why does increasing the temperature increase the rate of a chemical reaction?
2 Why does increasing concentration increase the rate of a chemical reaction?
3 Why does adding a catalyst increase the rate of a chemical reaction?

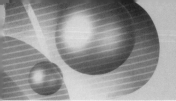

2.20 Equilibria and their importance in industrial processes

Reversible reactions and dynamic equilibria

Many reactions are **reversible**; they can go forwards or backwards.
For example

$$SO_2(g) + \tfrac{1}{2}O_2(g) \rightleftharpoons SO_3(g)$$

- The $\rightleftharpoons$ sign means that the reaction is reversible.
- If the forward reaction is exothermic the enthalpy change will have a negative sign. The reverse reaction will be endothermic and the enthalpy change will have a positive sign.
- Both reactants and products are constantly being made and broken up.
- If a reversible reaction takes place in a **'closed system'** where nothing can enter or leave then an **equilibrium** can be reached.
- At equilibrium there is a balance between the reactants and the products and the concentrations of the products and reactants stay the same. This does not mean that the concentration of reactants is equal to the concentration of the products.
- The forward and backward reactions have not stopped. It is just that at equilibrium the rate of forward reaction is equal to the rate of backward reaction.
    - As a result the equilibrium is described as being dynamic.

Le Chatelier's principle

Le Chatelier's principle helps us to predict the effect of changing factors that affect the position of equilibria.

The effect of catalysts on the position of equilibrium

A catalyst is a chemical which speeds up the rate of a chemical reaction but is not itself used up. This means that catalysts can be used many times. Transition metals and transition metal compounds are often good catalysts. Catalysts are widely used in industry to reduce the costs of producing chemicals.

- Catalysts increase the rate of chemical reaction by offering an alternative reaction pathway with a lower activation energy.
- Catalysts do not affect the position of equilibrium.

The effect of changing temperature on the position of equilibrium

If a reaction is at equilibrium then changing the temperature may affect the position of equilibrium.

- The position of equilibrium will shift to minimize the change in temperature.
    - As a result if the temperature is increased the position of equilibrium will shift in the endothermic direction.

Example

The reaction between A and B to form C is exothermic A + B $\rightleftharpoons$ C

If the temperature is increased then the position of equilibrium will shift in the endothermic direction. At equilibrium there will be a lower yield of C.

- As a result if the temperature is decreased the position of equilibrium will shift in the exothermic direction.

Example

The reaction between X and Y to form Z is endothermic X + Y $\rightleftharpoons$ Z

If the temperature is decreased then the position of equilibrium will shift in the exothermic direction. At equilibrium there will be a lower yield of Z.

The effect of changing the pressure on the position of equilibrium

If a reaction which involves gases is at equilibrium then changing the pressure may affect the position of equilibrium.

- The position of equilibrium shifts to minimize the change.
  - If the pressure is increased it will shift the position of equilibrium towards the side which has fewer gas molecules.
  - If the pressure is decreased it will shift the position towards the side with more gas molecules.

The effect of changing the concentration on the position of equilibrium

If a reaction involving solutions is at equilibrium then changing the **concentration** will affect the position of equilibrium.

- The position of equilibrium shifts to minimize the change.
  - If the concentration of one of the reactants is increased it will shift the position of equilibrium towards the right so more product will be made.
  - If the concentration of one of the reactants is decreased it will shift the position of equilibrium towards the left so fewer products will be made.

Equilibria and chemical processes

The Haber Process

Ammonia is used to produce fertilizers and explosives.

Ammonia, NH_3, is produced by the Haber Process. The reaction is reversible and the forward reaction is exothermic.

$$N_2(g) + 3H_2(g) \rightleftharpoons 2NH_3(g)$$

- An iron catalyst is used to increase the rate of reaction. This allows us to carry out the reaction at a reasonable temperature.
- The catalyst does not affect the position of equilibrium, so it has no effect on the yield of ammonia at equilibrium.
- The forward reaction between nitrogen and hydrogen is exothermic. Increasing the temperature increases the rate of reaction but decreases the yield of ammonia (increasing the temperature shifts the position of equilibrium in the endothermic direction).
  - As a result a compromise temperature is used. This gives a reasonable rate and a reasonable yield of ammonia.
- Increasing the pressure increases the yield of ammonia (increasing the pressure shifts the position of equilibrium towards the products side, which has fewer gas particles).

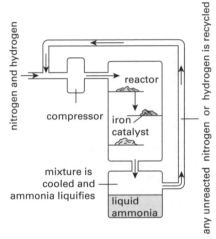

The Haber Process

The contact process

Sulfuric acid is produced by the contact process. The reaction is reversible and the forward reaction is exothermic.

$$SO_2(g) + \tfrac{1}{2}O_2(g) \rightleftharpoons SO_3(g)$$

- A vanadium(V) oxide catalyst is used to increase the rate of reaction. This allows us to carry out the reaction at a reasonable temperature.
- The catalyst does not affect the position of equilibrium, so it has no effect on the yield of sulfur trioxide at equilibrium.
- The forward reaction between sulfur dioxide and oxygen is exothermic. Increasing the temperature increases the rate of reaction but decreases the yield of sulfur trioxide (increasing the temperature shifts the position of equilibrium in the endothermic direction).
  - As a result a compromise temperature is used. This gives a reasonable rate and a reasonable yield of sulfur trioxide.
- Increasing the pressure increases the yield of sulfur trioxide (increasing the pressure shifts the position of equilibrium towards the products side, which has fewer gas particles).
- In practice a high yield is obtained without a high pressure. As maintaining a high pressure is very expensive, a lower pressure is used.

Questions

1 What does the sign $\rightleftharpoons$ mean?

2 What does Le Chatelier's principle state?

3 How does adding a catalyst affect the position of equilibrium?

The periodic table of the elements

Key

| relative atomic mass |
| **atomic number** (symbol) |
| name |
| atomic (proton) number |

Example: 1.0 **H** hydrogen 1

(1)	(2)	(3)	(4)	(5)	(6)	(7)	(8)	(9)	(10)	(11)	(12)	(13)	(14)	(15)	(16)	(17)	(18) / 0
1.0 **H** hydrogen 1																	4.0 **He** helium 2
6.9 **Li** lithium 3	9.0 **Be** beryllium 4											10.8 **B** boron 5	12.0 **C** carbon 6	14.0 **N** nitrogen 7	16.0 **O** oxygen 8	19.0 **F** fluorine 9	20.2 **Ne** neon 10
23.0 **Na** sodium 11	24.3 **Mg** magnesium 12											27.0 **Al** aluminium 13	28.1 **Si** silicon 14	31.0 **P** phosphorus 15	32.1 **S** sulphur 16	35.5 **Cl** chlorine 17	39.9 **Ar** argon 18
39.1 **K** potassium 19	40.1 **Ca** calcium 20	45.0 **Sc** scandium 21	47.9 **Ti** titanium 22	50.9 **V** vandium 23	52.0 **Cr** chromium 24	54.9 **Mn** manganese 25	55.8 **Fe** iron 26	58.9 **Co** cobalt 27	58.7 **Ni** nickel 28	63.5 **Cu** copper 29	65.4 **Zn** zinc 30	69.7 **Ga** gallium 31	72.6 **Ge** germanium 32	74.9 **As** arsenic 33	79.0 **Se** selenium 34	79.9 **Br** bromine 35	83.8 **Kr** krypton 36
85.5 **Rb** rubidium 37	87.6 **Sr** strontium 38	88.9 **Y** yttrium 39	91.2 **Zr** zirconium 40	92.9 **Nb** niobium 41	95.9 **Mo** molybdenum 42	98.9 **Tc** technetium 43	101.1 **Ru** ruthenium 44	102.9 **Rh** rhodium 45	106.4 **Pd** palladium 46	107.9 **Ag** silver 47	112.4 **Cd** cadmium 48	114.8 **In** indium 49	118.7 **Sn** tin 50	121.8 **Sb** antimony 51	127.6 **Te** tellurium 52	126.9 **I** iodine 53	131.3 **Xe** xenon 54
132.9 **Cs** caesium 55	137.3 **Ba** barium 56	138.9 **La*** lanthanum 57	178.5 **Hf** hafnium 72	180.9 **Ta** tantalum 73	183.9 **W** tungsten 74	186.2 **Re** rhenium 75	190.2 **Os** osmium 76	192.2 **Ir** iridium 77	195.1 **Pt** platinum 78	197.0 **Au** gold 79	200.6 **Hg** mercury 80	204.4 **Tl** thallium 81	207.2 **Pb** lead 82	209.0 **Bi** bismuth 83	210.0 **Po** polonium 84	210.0 **At** astatine 85	222.0 **Rn** radon 86
[223.0] **Fr** francium 87	[226.0] **Ra** radium 88	[227] **Ac†** actinium 89	[261] **Rf** rutherfordium 104	[262] **Db** dubnium 105	[266] **Sg** seaborgium 106	[264] **Bh** bohnium 107	[277] **Hs** hassium 108	[268] **Mt** meitnerium 109	[271] **Ds** darmstadium 110	[272] **Rg** roentgenium 111							

Elements with atomic numbers 112–116 have been reported but not fully authenticad

* 58 – 71 Lanthanides

140.1 **Ce** cerium 58	140.9 **Pr** praseodymium 59	144.2 **Nd** neodymium 60	144.9 **Pm** promethium 61	150.4 **Sm** samarium 62	152.0 **Eu** europium 63	157.3 **Gd** gadolinium 64	158.9 **Tb** terbium 65	162.5 **Dy** dysprosium 66	164.9 **Ho** holmium 67	167.3 **Er** erbium 68	168.9 **Tm** thulium 69	173.0 **Yb** ytterbium 70	175.0 **Lu** lutetium 71

† 90 – 103 Actinides

232.0 **Th** thorium 90	231.0 **Pa** protactinium 91	238.0 **U** uranium 92	237.0 **Np** neptunium 93	239.1 **Pu** plutonium 94	243.1 **Am** americium 95	247.1 **Cm** curium 96	247.1 **Bk** berkelium 97	252.1 **Cf** californium 98	[252] **Es** einsteinium 99	[257] **Fm** fermium 100	[258] **Md** mendelevium 101	[259] **No** nobelium 102	[260] **Lr** lawrencium 103

Index